VIBRATION

VIBRATION

Based on six lectures delivered at
The Royal Institution, London
in December 1962 by

R. E. D. BISHOP

KENNEDY RESEARCH PROFESSOR IN THE UNIVERSITY OF LONDON AND FELLOW OF
UNIVERSITY COLLEGE

SECOND EDITION

CAMBRIDGE UNIVERSITY PRESS

CAMBRIDGE

LONDON · NEW YORK · MELBOURNE

CAMBRIDGE UNIVERSITY PRESS
Cambridge, New York, Melbourne, Madrid, Cape Town, Singapore, São Paulo, Delhi

Cambridge University Press
The Edinburgh Building, Cambridge CB2 8RU, UK

Published in the United States of America by Cambridge University Press, New York

www.cambridge.org
Information on this title: www.cambridge.org/9780521296397

First published 1965
Second edition 1979
Re-issued in this digitally printed version 2009

A catalogue record for this publication is available from the British Library

ISBN 978-0-521-22779-7 hardback
ISBN 978-0-521-29639-7 paperback

Contents

Preface to the first edition

But though the compliment implied
Inflates me with legitimate pride,
It nevertheless can't be denied
That it has its inconvenient side.

An invitation to give the Christmas Lectures at the Royal Institution is hard to resist. These are not the sort of lectures that a university teacher normally inflicts. Being mainly for children, they are fun to give; and there is a tradition – this was to be the 133rd series – that the lectures contain many illustrations and demonstrations.*

The first thing was to find a subject, and 'Vibration' seemed as good as any. Vibration is essentially a practical matter for engineers and the lectures could be based entirely on simple demonstrations and experiments. But in writing out (or, rather, around) what I actually said, I have experienced some difficulty. The printed word is very different from the spoken and mere description of some experiments is a poor substitute for actually performing them. It is one thing to say that a dog can hear sound waves at ultrasonic frequencies and quite another to show it with Meg, an enthusiastic bull mastiff.

Why bother to write the book then? Well the reason is this – engineers have made far too little effort in the past to explain

---

* My assistants and I showed about one every two minutes, on average, for some 7½ hours in all; that is just about par for the course.

to others what their profession is all about. An engineer derives satisfaction from making things and making them function properly. He attempts to steer a course between the hazards of sloppy thinking on the one hand and, on the other, such unswerving devotion to scientific rigour as to display a sort of rigor mortis. By describing one fairly specialized – though vitally important – side of mechanical engineering, this little book will, I hope, illustrate what all this can imply.

Many people have helped in the preparation, both of this book and of the lectures upon which it is based. What with typing, making drawings, photographing, demonstrating, criticizing the manuscript, making pieces of apparatus, lending equipment and films, and so forth, these friends have immensely simplified things for me. To all of them, I offer my thanks.

R.E.D.B.

Preface to the second edition

Let us grasp the situation,
Solve the complicated plot
Quiet calm deliberation
Disentangles every knot.

Judging from the enormous literature on the subject, a book on vibration which contains no mathematics and which is based on simple experiments may be thought something of a curiosity. Perhaps it is. Even so I am not persuaded that there is much point in making this little book more relentless and, accordingly, I have retained its descriptive approach.

This edition differs from the first in that I have added a couple of new sections. One is on the behaviour of ships in waves and its main purposes are to give some idea of the sort of thing that a research worker in vibration analysis might try to sort out and to place more emphasis on non-conservative systems. The other new section is on the elements of transient vibration. I have also made a number of smaller modifications in the text, largely with a view to bringing it more up to date.

All the quotations used in this edition have been taken from that most English of men, W. S. Gilbert.

R.E.D.B.

‖‖‖

Vibration: friend or foe

The moon in her phases is found,
The time and the wind and the weather;
The months in succession come round
And you don't find two Mondays together.

In 1807 the celebrated Dr Thomas Young published the lectures he gave in the Royal Institution. He complained that, in his day, Sound and Vibration were '. . . treated in a very abstruse and confused manner, or connected entirely with the practice of music, and habitually associated with ideas of mere amusement'. We must clearly take our subject very seriously. But, before starting off let us take heart from another of Thomas Young's observations – namely that '. . . . many of the phenomena belonging to (the theory of sound and vibration) are so remarkable and so amusing, as amply to repay the labour of examining them by the entertainment that they afford'. Perhaps, after all, we shall be able to look forward to some 'mere amusement'.

1.1 The familiarity of vibration

A moment's thought will reveal several possible reasons why Thomas Young was so forthright. He was a leading physician as well as physicist; he may, therefore, have had the human body in mind. After all, our hearts beat, our lungs oscillate, we shiver when we are cold, we sometimes snore, we can hear and speak because our eardrums and our larynges vibrate. The light waves which permit us to see entail vibration. We move by oscillating our legs. We cannot even say 'vibration' properly without the tip of the tongue oscillating. And the matter does not end there – far from it. Even the atoms of which we are constituted vibrate.

It may seem rather indiscriminate to lump cold shivers and heart beats on the one hand with the vibrations of light and of atoms on the other. It is not obvious where to draw the line and, in fact, it is sometimes difficult even to decide what is a vibration and what is not. Are the tides a form of vibration, for example? Pursuit of this point is unlikely to be fruitful so let us merely observe that, if we are prepared to stretch the definition of 'vibration' a little, it quickly becomes apparent that many of the events of everyday life have an extraordinarily cyclic quality. It is a curiously shaky world we live in.

It is no exaggeration to say that it is unlikely that there is any branch of science in which vibration does not play an important role. Our purpose in this book will not be to discuss the phenomenon in all its generality, however, but rather to discuss it as something with which engineers have to contend. Let us limit ourselves to *mechanical* vibration so that even such familiar things as oscillatory currents in electric circuits or the fluctuating temperature of a hot-water system will be excluded. It is still a pretty sizable subject, so let us limit our discussion further by imposing three more restrictions. To begin with, few matters of technique – either experimental (as in vibration measurement) or theoretical – need be raised. Secondly, let us, as far as possible, leave out discussion of related subjects – such as sound, noise and the physiological side of vibration; these are all fascinating and technically important, but it would make this book too diffuse if we were to dwell on them at any length. Finally, it will be necessary for us to omit many details, and even some explanations. This will be quite unavoidable when we come across topics that are the subject of active research. But even when this is not so, there seems little point in devoting a substantial part of a whole chapter to a single problem.

It may seem rather futile to waste much time on a simple phenomenon like vibration, for after all it is only an 'in-and-out' motion. To be sure, we can all think of instances in which this shaking motion is superimposed on some bodily motion – as with the flapping wings of a bird in flight. Even so, would it not be fair to say that a study of vibration is likely to be pretty dull by reason of its narrowness? This is a fair question, and the answer would undoubtedly be 'yes' if we were only to study the motion, without reference to what causes it.

But if we concern ourselves with the reasons *why* oscillation occurs, we are often confronted by interesting – and sometimes surprising – behaviour. Nothing could be simpler than an oscillating pendulum and we might expect to understand its motion without much difficulty. A motor-car shakes because it runs over a rough road and because its engine is running; matters are now becoming more complicated, though not much more subtle. When the button of an electric bell is pressed the clapper starts moving backwards and forwards (even though the bell may be driven from a d.c. battery); evidently the fluctuating force that moves the clapper does not exist when the clapper remains at rest. Although there is no *externally applied* oscillatory disturbance, there is an oscillatory motion, so things are becoming a little more interesting. (Although the electric bell seems unlikely to provide much of an obstacle to our understanding, it turns out that the principle upon which it works is not as simple as it might seem at first sight.)

These are not very exciting examples perhaps. Let us consider one that was. Fig. 1 shows a picture taken during a famous oscillation – that of the great Tacoma Narrows Bridge in Washington State. It was caused by a *steady* wind and led to the ultimate destruction of this fine structure only a few months after its completion. Now this particular vibration was obviously not intended and for some time the reason for its onset was not understood. No structural engineer would want to repeat this costly mistake, so it has been the object of very close scrutiny.

Few vibrations are as spectacular as that of the Tacoma Bridge, but there have been other remarkable displays and the picture shown as fig. 2 was taken during one of them. A few years ago some engineers in this country had the idea that oil could be transported cheaply by sea in 'nylon oil barges'. Experience has shown that it is indeed a commercially acceptable proposition. The oil is pumped into a long sausage-shaped nylon bag which floats in the sea water because of the relatively low specific gravity of oil. Only the top of the bag shows above the surface of the water. A tug tows this 'barge' in the direction of its longitudinal axis. It all sounds rather simple, but several interesting and unusual design problems had to be solved before success was achieved. One of the

Fig. 1. The suspension bridge over the Tacoma Narrows in Washington State. Soon after it had been opened, the bridge swayed violently in the wind and was destroyed. (Courtesy B. D. Eliott.)

most interesting of these was a vibration problem. For it was discovered that, unless proper precautions are taken, the barge (or 'dracone' as it is sometimes called) executes enormous 'snaking' oscillations: instead of following its tug with a modest docility, it prefers to follow a wild, zig-zag course.

If the Tacoma Bridge and the dracone seem sufficiently strange and remote to be almost curiosities, let us look at yet another spectacular failure. Fig. 3 shows half a ship; it is in fact the after end of the tanker *Pine Ridge* which broke in two during a storm in the Western Atlantic during December 1960. The waves that she endured during her useful life (and particularly those in that last fatal storm) set up stresses in the hull. In the end the steel gave up the struggle and the ship broke in two. Here, at least, one would think that the issues are fairly clearly understood and that a ship is unlikely to be lost in this way nowadays. But in reality things are distinctly unsatisfactory; accurate estimation of hull stress is very

Fig. 2. A nylon oil barge zig-zagging across the path of its tug. (Courtesy Dracone Developments Ltd.)

difficult even without the metallurgical problem of deciding what stress is safe. Actually the present position is essentially one of ignorance. While the vast majority of very large ships serve a useful life of perhaps 20–25 years and are then broken up, a few suffer serious structural failure and founder for reasons that are not at all clear.

We are beginning to see therefore that the study of vibration sometimes becomes very complicated and quite interesting. But unfortunately it is a subject in which the stakes may be high; it can be a matter of life or death.

1.2 The engineer's attitude to vibration

Large sums of money are spent nowadays on the study of various forms of vibration. Sometimes the object is to control it as being something that is fundamentally desirable. More often, the object is to find the reason why oscillation is set up and if possible to stop it.

Fig. 3. The tanker *Pine Ridge* which broke in two off Cape Hatteras in the Western Atlantic during December 1960. (Courtesy United Press International (U.K.) Ltd.)

Let us briefly consider some of the things that an engineer takes into consideration in deciding whether or not vibration matters in a particular project.

As we have already seen, mechanical agitation is not always just a troublesome by-product of engineering practice. On the contrary, it is often useful and may be essential. Occasionally, for example, an engineer finds it necessary to draw the cork from a bottle. If it is tight, he will do what anyone else would – reduce the friction force opposing withdrawal by twisting the cork back and forth. He employs an undeniably useful oscillation.

There are washing machines on the market which rely upon agitation for their functioning, and there are many other examples of mechanical shakers for mixing things up; a dentist, for instance, may use a special mechanical shaker for mixing amalgam. Alternatively, vibration can be used to *un*mix things, as in sieves and other sorting devices. Concrete will flow far more readily into the furthermost recesses when it is poured into shuttering if it is suitably vibrated with a probe; this is standard practice in structural engineering.

Many useful vibrations are not associated with 'agitation' at all, as in clocks, watches and metronomes. Again, one way of conveying grain from one place to another is to make it jump there along a vibrating conveyor.

Sometimes vibration is employed in medical practice. Machines are made, for instance, whose purpose is to massage away patients' unwanted bulges. Again, vibration of very high frequency has been found to have many uses – some of them rather surprising. Thus a dentist who really wanted to, could drill a square or a triangular hole in a tooth by means of vibration.

When we turn to the objectionable features of vibration, we find that, as usual, the human body presents a number of problems. Engineers go to considerable lengths in trying to keep the human body in its normal state. In terms of vibration, this form of endeavour ranges all the way from the manufacture of valves for the heart to the prevention of rolling in ships at sea.

In recent years there has been a tremendous interest in high-fidelity sound reproduction – a less serious form of pandering to human weakness. Engineers are constantly working to produce better recording and transmitting apparatus. This illustrates a much wider truth, namely that the whole subject of communication opens

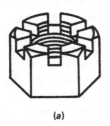 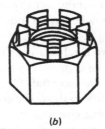 

(a)   (b)   (c)

Fig. 4

up plenty of vibration problems.

Engineers are sometimes worried about vibration because it may cause 'mal-functioning' of apparatus. An example of this can be found in a motor-car. At some engine speeds the rear-view mirror will oscillate so that a blurred image is seen in it. Electronic apparatus which is carried in aircraft or missiles has often to be supported on 'anti-vibration' mountings in order that the shaking of its environment shall not affect it. Again, a vibrating cutting tool will prevent a turner from producing accurate, and well-finished work with a lathe. Other examples will no doubt suggest themselves.

If a nut is tightened on a bolt which withstands a fluctuating load, it may work loose. Quite simple tests show that even the tightest nut can be loosened by a judicious vibration – often in a matter of a few seconds. For this reason slotted and castellated nuts (see fig. 4(a) and (b)) are used in engines for such vital purposes as holding down bearing caps, split pins being inserted to lock them. There are, in fact, several devices available for preventing nuts from loosening and fig. 4(c) shows one type of lock-nut. Again, turn-buckles that are used in the standing rigging of boats should always be locked by some means. One method is to tie them with wire in such a way that any tendency to slacken puts the locking wire in tension. An engine fitter or a boat rigger has to pay great attention to details like this, as the opening up of a bearing or the parting of a shroud could be serious.

The correct performance of many machines depends on the accurate geometry of their working parts. For instance, accuracy of the tooth profiles of gear wheels may be essential. If vibration occurs, it may greatly increase the rate at which parts become worn and so diminish the performance of the machine concerned. The rotating shutter of a high-speed camera is sometimes driven through a gear train and if this assembly executes a vibration while it is running the gear teeth may be ground unmercifully and all precision of the shutter operation will then be lost. It may be remarked here that small errors in the geometry of parts – and gear wheels are a case in point – may *cause* vibration.

A vibration is associated with fluctuating stresses and, during a sufficiently violent motion, these may become large enough for breakage to occur. That is what happened with the Tacoma Bridge (fig. 1). This effect can be demonstrated by throwing a piece of brittle material into a sufficiently violent oscillation. Thus, a glass tube may be made to execute longitudinal vibrations (like a concertina) if it is rubbed strongly along its length with a damp cloth, and this may cause it to break.

The breakage of a component during a very violent vibration is understandable, even if it is untoward. But unfortunately it is not the only way in which failure can occur. Undoubtedly the worst feature of vibration is that it can cause fatigue of metals, reinforced plastics or other structural material. This type of failure is usually as drastic as it is unexpected and it often amounts to treachery in one of its purest forms, since there is usually no warning: some component which has been vibrating over an extended period of time suddenly snaps. Despite prolonged investigation, the precise reason for this has not yet been discovered.

The failure of a piece of metal can easily be demonstrated in a crude fashion by bending a length of metal strip first one way and then the other until it parts. The strip can easily withstand a few bends, but gives up after a while.

The phenomenon of fatigue failure is associated with high local stresses and it is surprising how often these high stresses seem to be unavoidable. Fatigue can occur for instance at a point on a surface over which a succession of heavily loaded balls has constantly to pass, as in a ball bearing; the passage of each ball gives rise to a cycle of stress so that the problem is, at least in a sense, one of vibration.

As a matter of fact, the possible onset of fatigue is a more important reason than the reduction of machining accuracy for trying to prevent vibration of lathes and other machine tools. Modern cutters are capable of removing metal rapidly, but their

 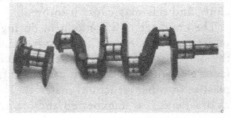

Fig. 5. Examples of fatigue failure. (a) A 5 cm diameter shaft broken by a twisting oscillation superimposed on high-speed rotation. (b) Valve spring from a motor-car engine. (c) Crankshaft from a motor-car engine.

hardness and wear-resistance have been achieved only at the expense of resistance to fatigue.

Fig. 5 shows some examples of fatigue failure. Picture (a) is of a shaft which broke as a result of torsional (twisting) vibration, (b) shows a broken valve spring from a car engine, (c) shows a break in the crankshaft of a motor-car. These three failures were all caused by vibrations. If we were to examine the breaks carefully, we should find that, in each, the metal has a characteristic 'grainy' appearance. The fact that fatigue is initiated in regions of high stress is well illustrated in fig. 5(a), for the fracture in that case was started in the corner of a keyway – a familiar source of high local stress.

The components shown in fig. 5 all broke quickly and cleanly. This is a feature of fatigue failure and one reason why its effects can be devastating. Fig. 6(*a*) shows the bladed rotor of an axial-flow air compressor from a jet engine; in operation it rotates at about 10 000 revolutions per min. Now the rush of air through the compressor may set one or more of the blades in vibration and if failure occurs the snapped blade will rattle round and may very well ruin the rest of the blades, as shown in fig. 6(*b*). The prevention of blade vibration in gas turbines is therefore an extremely serious matter.

The results of some vibrations are both dramatic and very serious. But these results are not necessarily commonplace. Broken crankshafts were plentiful before World War II but are comparatively rare nowadays, at least in cars that have travelled less than 200 000 km. That problem has been solved. Again compressor blading is *known* to raise difficulties so that precautions are made to ensure that the type of catastrophe illustrated in fig. 6(*b*) does not happen in service. New departures in engineering, whether they relate to radically new designs or to improvements (in any sense) in the performance of machines, *may* bring attendant vibration problems; but, provided they are of a recognized kind, these would only catch the unwary. Generally speaking, safety is only prejudiced when some previously unrecognized oscillation causes failure – as with the Tacoma Bridge.

Occasionally – *very* occasionally – a potentially dangerous oscillation of a new kind is spotted before it can occur. This has been the case, for instance, with tower buoys. In its simplest form a tower buoy is a huge pencil-like underwater structure that is attached to the sea bed by means of a universal joint as shown in fig. 7; it supports a pipe which is led to it along the bottom and which conveys crude oil. The oil flows up the column-supported pipe and can be discharged into a tanker that is moored to the top of the column. It has been pointed out that, in theory at least, conditions could arise in which the column would perform dangerous *bending* oscillations in rough sea.

Some very serious and familiar problems, by contrast, were recognized as vibration problems long ago but proved too difficult to solve. Thus it was clearly pointed out in the 1920s that a ship is an elastic structure which suffers an oscillatory distortion in heavy seas. It is obviously very desirable to have reliable design rules by which hulls can be designed, but in those days the means of drawing up such rules simply did not exist. A great deal of progress

Fig. 6. The rotor of an aircraft jet engine compressor. (a) The rotor as installed. (b) The result of blade failure due to fatigue.

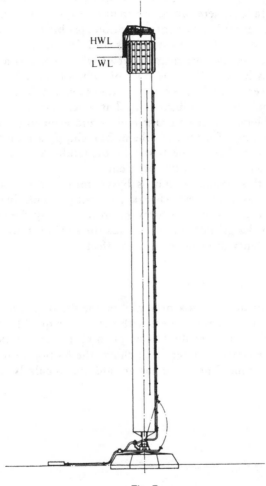

HWL
LWL

Fig. 7

has since been made in structural theory, in the study of fluid flow, in statistical techniques, in oceanography and in computing. Now it is necessary carefully to discard the more empirical rules that have been developed over many years in favour of the far more powerful techniques of vibration analysis that are becoming available.

The distortions of tower buoys and of ships are still subjects of debate among specialists and research workers and so are rather special cases. In fact there can be few major industries which do not have well-recognized types of vibration problem. Competent engineers are very vibration-conscious and few *serious* types of breakage or loss of performance occur of which there is anything like a previous history. But this state of affairs is only bought by continuous research and development and the expenditure of large sums of money. Some industries find it essential to keep their common problems under constant review and even to maintain a central organization for tackling them. The shipping industry, for instance, finds this to be the best way of handling its peculiarly intractable machinery vibration problems.

Despite all this, fatigue failure is by no means rare. It is quite common in components whose breakage is not particularly serious, simply because the expense of prevention (if only by the employment of better designers) would be unjustified. Almost any garage can produce plenty of evidence to this effect.

1.3 The nature of vibration

The dullest part of this book may well be the discussion of what a vibration *is*. Let us therefore deal with the matter quickly, observing, first of all, that basically it is simply a to-and-fro motion.

Fig. 8 is an electro-cardiogram; it shows the motion of my pulse plotted against time. The curve has an odd, but regularly repeated

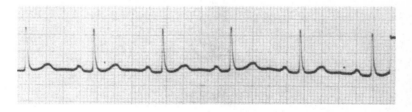

Fig. 8

shape, and one cycle of events takes approximately 0.78 s. The 'period' of the pulse is thus 0.78 s, though an engineer would be tempted to look at the matter another way and say that, since 1.28 cycles take place in a second, the oscillation has a 'frequency' of

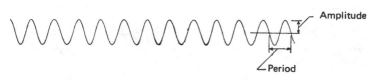

Fig. 9

1.28 cycles per second.* Frequency is the most important quantity that an engineer needs to know in any practical vibration problem.

It would obviously be difficult to compare the curve in fig. 8 with that of fig. 9 which represents the motion of a point in a particular wooden beam that was being shaken by an electrical device during a certain experiment. This is because the shapes or 'wave-forms' of the two curves are so very different. Now it happens that the shape of any regularly repeated curve like that of fig. 8 can be built up from a suitable assembly of sine curves. The curve in fig. 9 is an almost pure sine curve. The sine waveform thus begins to assume a particular importance and we shall need to be able to specify a sinusoidal motion. The maximum departure is called the 'amplitude' and the time for a complete cycle to be executed is, of course, the 'period' (see fig. 9).

Take the pulse motion of fig. 8 as an example. If the six sine curves in fig. 10(a) – each with its own frequency and amplitude – are added together, they produce the more complicated curve of fig. 10(b). By taking more and more of the components – and a very large number would be needed in this particular case – the latter curve could be made more and more like that of fig. 8 (indicated by the dotted line). The reverse process – of finding the sinusoidal components of a complex (but regularly repeated) curve – is called 'harmonic analysis', and it is of considerable technical importance since the components often have more to tell us than the composite curve. Actual analysis is usually performed, nowadays, by electronic 'wave analysers'.

A more important example of wave analysis, and one that is more rapidly accomplished, is shown in fig. 11. Diagram (a) shows a 'square' waveform while (b) shows its first three components. If

* Indeed, if he is on his best behaviour he will probably refer to a frequency of 1.28 'hertz'. But why this newer unit (with its abbreviation 'Hz') is thought to be superior to the more expressive 'cycles per second' (or 'c/s') is a little obscure. Nevertheless we will behave ourselves and refer to hertz in future.

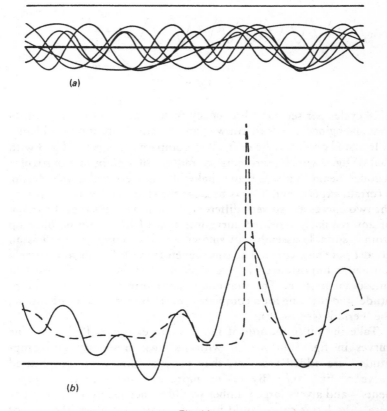

(a)

(b)

Fig. 10

these are added together, they alone give quite a fair approximation to the parent curve, as shown in (c).

This idea of harmonic analysis is a most convenient one since it suggests that the sine curve may be used as a basis for discussion of oscillatory quantities. Later on, therefore, we shall refer mainly to these simpler curves. If necessary we can build up more complicated ones from them.

There is no real necessity, at this stage, to go into the properties of sine curves in any detail. They are readily accessible in textbooks, and in any case it is not important that we should do more than remember these pleasingly clean-looking curves with their ability to combine to give more complex waveforms.

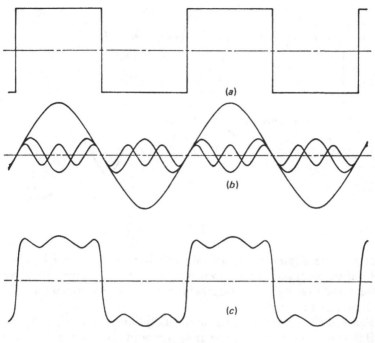

Fig. 11

It is usual and natural to think of vibration as a fluctuating motion or *displacement*. It will be realized, however, that, if we are to understand why the vibration takes place, then we shall have to think about fluctuating forces as well. All these definitions, then, relate just as much to curves of force, pressure, volume and so on, as they do to curves of displacement. We may speak of the amplitude and frequency of a pressure fluctuation, for instance.

Fig. 12 shows portions of two curves. Instead of trying to add them, let us suppose they represent two totally different things. In fact (a) is a curve of lung pressure and (b) is a corresponding curve showing volume of air inhaled. Naturally the two curves have the same period (and therefore frequency) since they represent two related phenomena; the frequency is about 0.13 Hz.

One of the curves is shifted horizontally relative to the other. This shift is called a 'phase difference'. Although a phase difference is represented by a distance measured along the time scale, and may

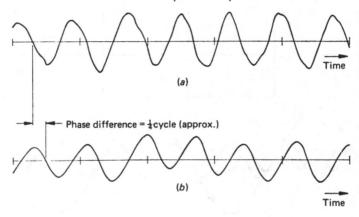

(a)

Phase difference = ¼ cycle (approx.)

(b)

Time

Fig. 12

therefore be expressed as an interval of time, it is much more helpful to compare this interval with the periodic time. Thus we should say that curve (b) lags behind curve (a) by approximately one quarter of a cycle.

Phase differences are sometimes vitally important. The electric bell that was mentioned earlier simply would not work if the force exerted by the electromagnet on the clapper were in phase with the displacement of the clapper. This fact is often missed in elementary explanations of how these bells work.

The crankshaft shown in fig. 5(c) broke because it twisted and untwisted excessively while it was rotating. This distortion was caused by the gas pressures in the cylinders and the mechanical effects of the reciprocating motions of the pistons and connecting rods. Crankshafts still execute this oscillatory motion as they rotate but they do not do it to excess because calculations now show how it can be limited. As one would expect, those calculations have to make allowance for the fact that the pistons are not all 'in step'; their relative motions are determined by the angular positions of the crank throws. Here, then, we have phase differences between the twisting efforts applied to the crankshaft.

Fig. 13(a) shows two sine waves of equal amplitude but with slightly different frequencies. Suppose that they represent two contributions to the same quantity so that they must be added together to give the total effect. This addition has been performed

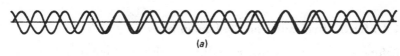

(a)

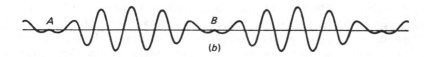

(b)

Fig. 13

in fig. 13(*b*). The total variation exhibits the phenomenon of 'beating'. The distance between the points *A* and *B* along the time scale represents the time which must elapse for the vibration having the higher frequency to make one more complete cycle than the one having the lower frequency. The smaller the difference between the frequencies of the two components, the longer will be the interval *AB* so that we have here the basis of a method for quite accurate measurement of small differences of frequency. It is on this principle that some chronometer adjustments are made, while the 'shapes' of vibrations are sometimes examined by flashing 'stroboscopic' lights which also operate on the principle of beating.

A fluctuating force may exhibit beating. Fig. 14 is a trace that was obtained in an experiment and it represents the variation of the

Fig. 14

vertical force which is exerted by a fluid flowing past a horizontal cylinder placed across its path, when that cylinder is moved bodily up and down sinusoidally. The beating waveform tells us that the force has two components whose frequencies are nearly equal; this was arranged by making the frequency of the up-and-down motion nearly the same as that which the wake would have had were the cylinder held fixed. For, as we shall see, an object that is placed in a fluid flow may set up a wake which swings from side to side like the tail of a goldfish.

1.4 Effects of vibration on the human body

Before we start discussing the nature of vibration and its problems in any detail we should perhaps know something about the ability of bodies to stand up to this motion. Two types of body that have to withstand vibration and have sometimes to be protected from it are particularly important, namely the human body and bits of metal. Now the question of just how much resistance these have to vibration opens up enormous fields of research, so that this discussion will have to be somewhat superficial.

We start with the human body. In the first place it is obvious that the human body as a whole can withstand very large changes of displacement indeed provided that they are performed slowly enough. It would be tedious to go up and down in a lift for very long, but it would not be painful. Within reason, then, amplitude of vibration alone is no problem. It is a matter of common experience, however, that, when the frequency begins to rise, things may become more serious. This will be confirmed by anyone who is prone to sea-sickness.

When an ocean liner pitches in a heavy sea, the first-class passengers (who live near the middle of the ship) may feel some discomfort as they are oscillated up and down. The tourist-class passengers (who live in the bows) oscillate at the same (pitching) frequency, but with a greater amplitude. For this reason alone, tourist travel can be less comfortable than first-class travel. In general, for a given frequency, the human body prefers smaller amplitudes (and first-class travel).

In *H.M.S. Pinafore* we meet a First Lord of the Admiralty named Sir Joseph Porter, K.C.B. He would certainly travel first class at sea, along with 'his sisters and his cousins and his aunts' who never seem to let him go anywhere alone. For one in his position, though, Sir Joseph is remarkably ill-informed about another aspect of sea-sickness. He says that he is perfectly content in a ship that is riding at anchor . . .

> But when the breezes blow
> I generally go below
> And seek the seclusion that a cabin grants

to which his adoring retinue adds

> And so do his sisters and his cousins and his aunts.

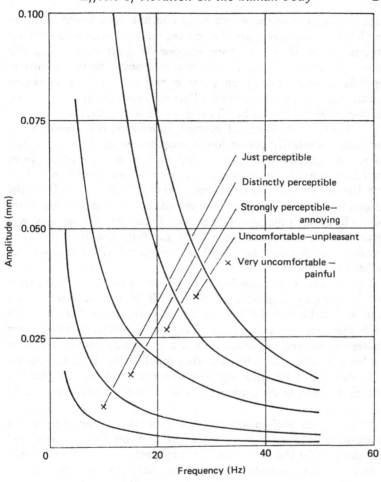

Fig. 15

The fact is that fresh air is the best antidote for feelings of sea-sickness whereas the hot air in a cabin can, by contrast, bring matters to a head very quickly. So here is another complication when large, slow oscillations are concerned; the proneness of a person to sickness may depend quite critically on atmospheric conditions.

Turning to vibration of smaller amplitude, the situation may be summed up roughly as in fig. 15. This set of curves relates to vertical vibration of the whole body and has been found useful in problems of road transport. The curves are approximate and could not possibly tell the whole story. For one thing there are perceptible differences between one person and the next; again, sensitivity depends to some extent on posture and on the way in which vibration is imparted to the body. Furthermore there is a psychological factor which has to be allowed for occasionally.

For a given amplitude of motion, moreover, the human body does not necessarily prefer lower frequencies as fig. 15 implies. There are certain 'bad frequencies' (which differ slightly from person to person). For example, a person who is seated in a chair that has a vertical vibration reacts violently at about 5 Hz; the distortion of the person is associated, at that frequency, with violent heaving of the 'shoulder girdle'. These 'bad frequencies' will become more understandable later on but, briefly, they exist because some parts of the body have comparatively little resistance to distortion.

A good deal of attention has to be paid by aircraft manufacturers and by railway engineers to all this. It is no use building an expensive airliner if the passengers are going to find it – or even *think* they find it – unbearably shaky. As a result of cooperation between engineers, physiologists and experimental psychologists, it has been possible to collect together much useful information for industrial use. Unfortunately a systematic study of all possible vibration effects in the human body would be a colossal undertaking.

Although the biological effects of a vibration depend upon the direction and site of its application and on its intensity and duration, by far the most significant characteristic of the vibration is its frequency. Straight away we may note that vibration in the range 18–18 000 Hz (approximately) is audible. The human ear is, in fact, a remarkably sensitive vibration-detecting device (though some of its characteristics are hard to explain). It can easily discriminate, for instance, between a sinusoidal vibration and a vibration of the same frequency but with a square waveform – the latter is more 'edgy'.

Table 1 gives some idea of the biological effects that may be expected when a human body is vibrated at various frequencies. The table is not intended to be taken very seriously, however, partly because of the indefiniteness that has already been mentioned in

Table 1

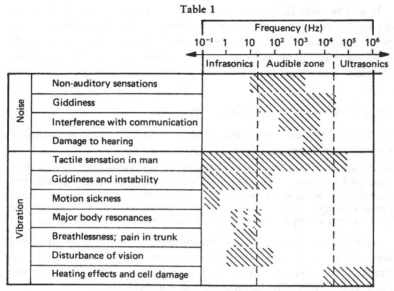

'Noise' on this chart means airborne vibration.

connection with fig. 15 and partly because – as if Mother Nature has no intention of letting us get away with anything – some biological effects are not clearly related to frequency at all; such effects have been omitted from the table.

It is, of course, necessary to know not only what the human body can withstand, but what it is likely to be asked to withstand. In an aeroplane, for instance, the sources of vibration may arise either from the propulsion system or from the aircraft structure and the atmosphere through which it is passing. Although no hard-and-fast rules could possibly be set down the following list will give some idea of what can be expected; in each case the source is known to have radiated vibration of sufficient intensity to disturb people in aircraft.

Piston engines and exhausts, 20–10 000 Hz.

Turbojets, 60–40 000 Hz.

Aerodynamic noise at high speeds, 150–40 000 Hz.

Turbulence and gusts, 0–5 Hz.

Vibration involving major structural distortion, 1–40 Hz.

The ram-jets and rockets that are used in aviation have caused disturbance to people over the entire audible range.

It will be clear from these remarks that it is not possible, here, to discuss the biological effects of vibration in detail. But if these notes give a rough idea of what happens when the human body is shaken, then we have achieved our aim and can turn our attention to metals.

1.5 The resistance of metals to vibration

Well over a hundred years ago, that observant Frenchman Stendhal wrote in his *Mémoires d'un Touriste:*

La Charité. 13 *April* 1837. I was passing at a good speed through the little town of La Charité, when, to punish me for my protracted thoughts this morning about the troubles to which iron is subject, the axle of my coach broke sharply ... I examined the grain of the iron of my axle; it had become coarse, apparently because it had been in use for a long time.

At that time, engineers were becoming aware that metal could sometimes 'wear out'; and, like Stendhal, they had noticed that breakage occurred suddenly and that the fractured parts had the curiously 'grainy' appearance seen in fig. 5.

This idea of iron 'wearing out' conflicted with the known behaviour of iron in stationary structures and attention became focused on this paradox by the apparently inexplicable failures of the axles used in railway rolling stock. These axles failed during service under loads that were appreciably lower than their known static breaking loads.

Fig. 16 shows an axle. The weight of the vehicle that it supports tends to bend it in the manner that is indicated; that is to say, the metal on the top surface becomes elongated and that at the bottom compressed. As the axle rotates once, a point on its surface experiences a cycle of tensile stress – no stress – compressive stress – no stress – and again tensile stress. This is illustrated in fig. 17.

After several initial investigations of this problem of railway axles had been made, a German engineer named Wöhler (who tested materials under the conditions of rotating bending that are illustrated in fig. 16), made the first systematic study of this phenomenon of fatigue. He found that if the alternating stresses were only slightly less than the static stresses which would cause breakage, only a few cycles of loading were required to cause failure. But as the alternating stress was reduced in amplitude the

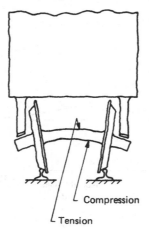

Compression

Tension

Fig. 16

number of cycles needed to cause failure increased. This tendency was maintained until the alternating stress level had been reduced to about a quarter or a third of the maximum sustainable static stress, at which level the life of the specimen appeared to be infinitely long. This limiting stress has become known as the 'endurance limit' of the material and the nature of the results is shown in fig. 18.

Fig. 3 shows a broken ship whose hull has been subjected to fluctuating stresses caused by waves. We noted that the stressing occurred in sea water, a corrosive liquid. Now although we know too little about the matter for comfort, it seems that the presence of the corrosive medium may substantially reduce the endurance limit of the steel.

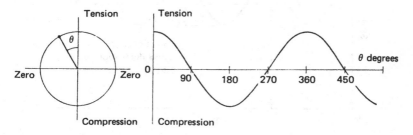

Fig. 17

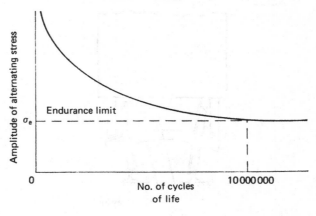

Fig. 18

In many engineering applications, material is not called upon to resist alternating tension and compression but has instead to resist a fluctuating stress superimposed upon a steady stress. This occurs, for example, in the wing spars of aircraft. The variation of stress as time goes on is somewhat like that shown in fig. 19. Under these circumstances it is found that, as the steady stress is increased, so the allowable alternating component for an infinite life is reduced. The curve of fig. 20 shows how the effects are related.

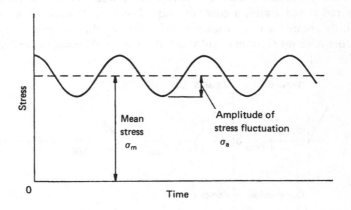

Fig. 19

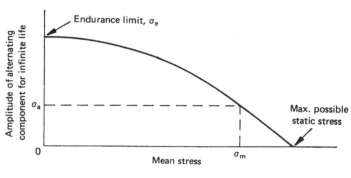

Fig. 20

Often the steady stress in a particular component is determined by the load to which it is subjected in service, while the alternating component arises from unwanted vibration in the system – as with the spar of an aircraft wing or the handle of a motorized garden implement (as in fig. 21). Clearly, when lightness and smallness are important criteria in the design, the mean stress level in the part must approach as closely as possible to the static strength and it is therefore of supreme importance that the alternating component due to vibration be kept as small as possible.

Fatigue is not really a feature *of vibration* as there is no necessity for the stress cycles to be regularly repeated; neither is the number of stress fluctuations in a given time important – at least under normal conditions. The point is that the number of stress cycles to cause failure of a component is usually large and the execution of a vibration is a common way of achieving the necessary large number in a relatively short time.

Although failures by fatigue may occur suddenly, it is not true to say that they occur without any warning whatsoever. Signs of damage become detectable as fatigue failure approaches – that is, provided the part concerned can be examined closely. The surface of the piece which is about to break becomes slightly cracked though the amount of warning of impending failure is very small if the stress fluctuation is large. Crack-detection is consequently a matter of great importance where fatigue is a possibility. An aircraft mechanic will always be on the alert for signs of hair-line cracks, and failure to report one would be a serious breach of discipline.

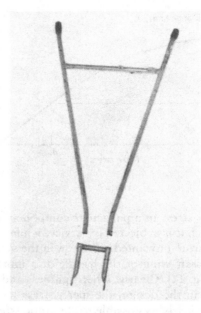

Fig. 21. The handle of a garden implement after a long hard life. Breakage occurred by fatigue. The steady bending stresses caused by downward pressure on the handle combined with fluctuating stresses set up by the engine and by the rotating blades to produce the failure.

1.6 Oscillation as a rigid body

Of the various forms of vibration, one is particularly suitable as an introduction. For want of a better word, let us call it 'shaking'. When a body undergoes this type of motion, it does not distort perceptibly.

The most obvious reason why a body should shake is simply that it is shaken by some external agency. Try to imagine a small boat resting on the surface of the sea. The boat will perform slow up-and-down motions as waves pass it – in the wake of a passing ship, for example. Since the boat does not creak as it performs this motion, it clearly does not distort very much. The motion can justifiably be referred to as 'rigid body oscillation'.

This particular type of 'rigid' motion has actually been *used*. A machine placed in the bows of a suitable small ship – like a tug – can be made to impart a 'rigid' pitching motion to the vessel.

If the ship is driven forward, it can be used as an efficient ice-breaker.

A human body that goes up and down in a lift also 'shakes'. The motion is not painful because it involves little distortion. Other examples of shaking will come to mind: the needle assembly of a sewing machine hardly distorts at all as it is moved up and down, neither do the pistons in a petrol engine, nor the trays in a sieve. In all of these examples, the body that shakes is not fixed rigidly to the ground and it shakes because some external device shakes it.

Later we shall discover the conditions under which oscillating bodies move as if they are sensibly rigid. We know from experience that such motion is possible, however, so let us briefly examine its consequences, since it is of both historical and technical significance. If the centre of mass of any body – or assemblage of bodies – oscillates, then Newton's laws tell us that this must be because a fluctuating force is being applied to the system by some agency. Looked at the other way round, this means that the system whose centre of mass is oscillating must be exerting a fluctuating force on its environment.

We are now concerned with what this conclusion implies in the special circumstances in which the system remains distortion-free. It presents no difficulties when the oscillating system is a body that is free to move and is simply shaken by some external means – as in the cases already mentioned. But that is not the only possibility, for a system may try to shake itself. It is not usually possible so to proportion the parts of a reciprocating engine that its centre of mass remains stationary while it is running. In other words, the act of starting an engine sets up a vibration, and fluctuating forces must act at the holding-down bolts.

This was, in fact, one of the first causes of trouble due to vibration in engineering. An early steam-engine could have a piston and piston rod weighing more than a tonne-f and, when these were accelerated to-and-fro along the cylinder, they produced a mighty heaving in the vicinity of the engine. So long as the operating speeds of these engines were low, the vibration could be kept at an acceptable level. But speeds could not be kept so low, and the inertia forces became a major nuisance as a consequence. It became necessary to 'balance' the engines, or, in other words, suitably to modify the distribution of the mass of the working parts so as to minimize the motion of the centre of mass.

Fig. 22. A piece of demonstration apparatus in which a short fat shaft is supported
in bearings that permit slight horizontal movement. The weights attached to the
surface of the shaft cause it to wobble when it is rotated by means of the central
driving belt.

The motion of the centre of mass may not be a necessary
consequence of the geometry of a system, even in the absence of
distortion. It may be an accident. This is familiar to those who dry
their washing in a spin-drier. If a load of washing is badly
distributed, it can cause the appliance to vibrate horribly. In this
motion, the spinner does not distort appreciably (though its
suspension will).

Consider a short fat shaft like that shown in fig. 22. It is not
possible to make it – or anything else for that matter – with
complete precision, and as a result the centre of mass of the shaft
will not lie exactly on its central axis. In fact the shaft in the
photograph was made as a demonstration model so that it has a
couple of offset screws to ensure that the centre of mass lies well

away from the central axis. The ends of this shaft are mounted in ball bearings and can run on horizontal guides so that the shaft is only restrained from vertical motion. By rolling from side to side the shaft can shake without distortion, and this is just what happens when it is rotated, the displacements at the two ends being in phase with each other. If the shaft were restrained and an attempt were made to hold its bearings still, then fluctuating forces would have to be applied to it.

If one of the two attached screws is removed from its hole and transferred to the diametrically opposite point on the surface, then this rolling motion takes place with the displacements at the two ends in 'antiphase' with each other; as one end moves one way the other moves the other way. This, again, is in keeping with Newton's laws, which tell us that a fluctuating rotation of a rigid body about its centre of mass requires an externally applied oscillatory torque. When the screws are on opposite sides of the short rigid rotor and rolling of the ends is permitted, the rotor does its best to keep its mass axis stationary with the result that its geometric axis wobbles.

To remove the need for the fluctuating forces which would be required to stop its shaking, a rotor must be balanced in a balancing machine and its mass axis adjusted to coincide with its geometric axis. Indeed, some rotors (such as those in gyroscopes) demand very careful balancing. Most of the rotors in common use, like that in a vacuum cleaner motor, are balanced fairly accurately before they are installed.

Generally speaking shaking motions present few problems. They are usually easy to identify and can often be sufficiently reduced by judicious balancing. As we shall discover, their essential feature is that they do not suffer a phenomenon known as 'resonance'. Unfortunately, mechanical vibration is a much more complicated phenomenon than this since distortion is *not* usually negligible, and so we must now start to think about distortions.

The ability to vibrate freely

Conceive me, if you can
An every-day young man;
A common-place type,
With a stick and a pipe
And a half-bred black and tan.

Pretty well every object possesses the ability to vibrate freely on its own after it has been disturbed. Although this 'free vibration' *as such* is not often of much interest in engineering, it is essential to have some understanding of it since, indirectly, it is vitally important. The point is that the behaviour displayed by a system in free vibration defines, in a very real sense, a sort of 'personality', and the dynamical personality of a system is what determines its behaviour under all sorts of conditions.

2.1 The nature of free vibration

After it has been struck by one of the hammers, a string of a piano is left vibrating on its own. It performs a 'free vibration', and it can do this because it has two properties. The first is that it has mass and is therefore capable of possessing kinetic energy by virtue of its motion. The second property is also a very common one; it has the ability to store energy by virtue of its distortion from its position of rest.

In the same way a simple pendulum can oscillate because (*a*) the bob is massive, and (*b*) it stores potential energy when the bob rises above its lowest position.

Fig. 23. The wing of an airliner. It is being bent by forces applied during a structural test. The wing is still well within its safety limit and will straighten out when the forces are removed. (Courtesy Vickers-Armstrong (Aircraft) Ltd.)

Later on, we shall touch on the oscillations of aeroplanes. Now it may seem that, while an aeroplane is massive all right, it is a rather rigid structure in which the storage of energy by distortion is unlikely. But in fact, to a dynamicist, an aeroplane seems sometimes to resemble a flying jelly. Fig. 23 shows the starboard wing of a VC 10 airliner during a static bending test (i.e. while the wing was not vibrating) and although the wing tip has been much displaced by the loading, the wing is still far from being damaged by this treatment. In just the same way, ships, houses, machines, people and things in general can all store some energy by distortion. And, since they all possess mass, this means that they can all vibrate freely if something starts them off.

It will best serve our purpose if we examine the free vibration of some very simple system. A length of bicycle chain suspended from one end will serve admirably because it will clearly display all the basic features of this motion.

If the chain is allowed to hang freely and come to rest, it may then be thrown into free vibration by displacing it sideways in some way and releasing it, or by striking it sharply. In starting the motion off, we should, logically, observe the rule that the sideways displacement at any point is 'small' – small, that is, in comparison with the length of the chain. The reason for this limitation need not distract us at this stage, and will become clearer in chapter 6. But it is hardly an oppressive restriction, because, after all, the amplitude of vibratory motion at a point in a structure is seldom comparable in magnitude with the dimensions of the structure itself when (as here) the vibration is one of distortion.

What happens to the chain? Although it is hard to describe the free vibration accurately, we can easily see that it has the following features:

(1) The time-history of the motion depends on how it is started.
(2) The motion dies away.
(3) There is no distinctive 'shape' of distortion of the chain during the motion and the vibration shape changes as time goes on. (Often the motion will finish up with a more or less recognizable shape and frequency.)
(4) It is quite impossible to distinguish a 'frequency' of oscillation of the chain. (One may emerge, however, as time goes on.)

So much for the free vibration that occurs after the chain has been released or struck. It seems chaotic, but, as we shall see, it can easily be sorted out.

Our first observation gives the clue as to how we might do this. If we are careful how we *start* the motion, we can completely change the chain's behaviour as regards observations (3) and (4); a steady shape *can* be obtained, together with a definite frequency of oscillation. The easiest way of showing this is to use what may at first seem to be a bit of sharp practice – though, in fact, it is perfectly legitimate.

If the top of the chain is mounted on a 'Scotch yoke' mechanism like that shown in fig. 24, the point of suspension can be oscillated from side to side with a sinusoidal displacement. If, too, the speed of the engine driving the mechanism can be controlled, the frequency of the oscillation can be varied. At very low frequencies, the chain simply shifts from side to side while still remaining more or

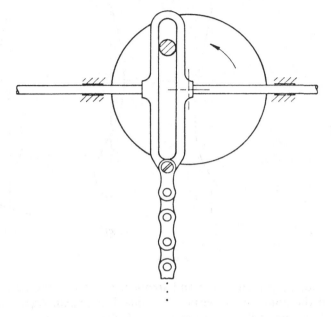

Fig. 24

less. vertical; this is actually the condition of 'shaking' described at the end of chapter 1. As the frequency is raised, however, there comes a stage at which the chain sways violently with the shape shown in fig. 25(*a*) and with the frequency of the driving motion. This is *not* a free vibration, as we are disturbing the chain, so we need not discuss yet why this large motion comes about. What is of interest is that, if the engine is now stopped suddenly, so that the support becomes fixed, a free vibration ensues – and this vibration is quite different from that started haphazardly. This free motion dies away, but accurately retains its shape and its frequency while it does so. In other words, the motion maintains its character as it subsides.

Is there any other frequency of the initial driving for which all this is true? This is a natural question to ask and it can easily be demonstrated that the answer is 'Yes – several'. With a somewhat higher frequency of the Scotch yoke, the chain can be made to execute quite large oscillations with the shape shown in fig. 25(*b*).

The ability to vibrate freely

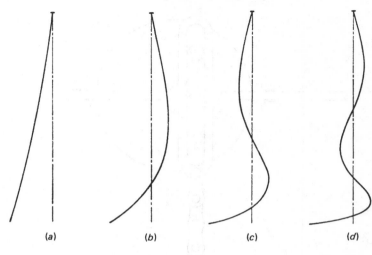

(a)   (b)   (c)   (d)

Fig. 25

These, too, keep their shape and frequency as they gradually die away, if the point of suspension is suddenly stopped. Again, at a higher frequency still, the shape of fig. 25(c) can be obtained. And so one can go on, though, with a given chain, the experiment gradually becomes more and more difficult to carry out convincingly.

The chain is thus shown to possess a series of shapes of vibration, each with an associated frequency. The shapes are known as *modes*. Each mode is associated with a rate of decay of motion, as well as with a frequency.

This more orderly way of looking at the chain's behaviour is not incompatible with what we see when we start the free vibration by striking the chain or by releasing it from some distorted form. The latter (complicated) motion is simply a mixture of motions in the various modes – each with its own frequency and rate of decay. The relative sizes of the contributions of the motions in the various modes to the total free vibration is determined by the way in which the total motion is started.

The human ear is very sensitive to the decay of free vibration when the frequency is in the audible range. This is actually one of the ways in which we distinguish one instrument from another. When a piano key is depressed, a hammer strikes the piano wires, which are then left vibrating freely; their motion then gradually dies away. If a tape-recording is made of this event, and is then played back *in the wrong direction*, the appropriate note is heard, but its intensity gradually increases until suddenly the sound vanishes. The recording does not sound like one of a piano.*

The frequencies of a system – any system, not just a hanging chain – its modes and its rates of decay are all, so to speak, part of the system and are not imposed on it by any external agency. They are so important that we shall now deal with them separately.

2.2 Frequencies of free vibration

The hanging bicycle chain has been seen to possess a series of characteristic frequencies. This is true of structures, machines and things in general. These systems can vibrate freely at one or more of those frequencies and, indeed, many of the free vibrations with which we are familiar are mixtures of this sort.

When A above middle C on the piano is struck, a note is heard which has the characteristic frequency of 440 Hz. This frequency is, really, only that of a *dominant* vibration as the piano strings do not only vibrate at 440 Hz, but also perform small additional vibrations in other modes with 2, 3, 4 . . . × 440 Hz. It is partly from these 'overtones' that we recognize that it is a piano and not, say, a bassoon that is being played.

The first few frequencies of a hanging bicycle chain turn out not to be evenly spaced apart like those of the piano strings. (Higher frequencies of the chain do gradually take up an even spacing, however.) In fact, this equality of spacing is not at all general. Thus successive antisymmetric modes of the VC 10 aircraft shown in fig. 23 were found to have frequencies of 1.85, 2.56, 2.92, 3.96, 4.28, . . . Hz. Indeed, there is no reason in fact why a pair of the frequencies should not be almost the same, or even be equal. This becomes obvious when one thinks of the oscillation of a conical pendulum in two perpendicular vertical planes, both passing through the point of suspension.

---

* The interested reader's attention is drawn to Haydn's Menuetto al Rovescio (from an unpublished Sonata in A major). The melody of this delightful piece of music is more or less the same backwards as forwards. If a tape-recording of it played on a piano is replayed backwards, the melody is quite recognizable although the music appears to be somewhat like that of a harmonium.

If a vibration can be started off accurately in one mode, it remains in that mode, but dies away. It retains the frequency of that mode while this happens. All this may be demonstrated with the chain. Let us now disregard the decay of such a single-mode motion and think only about the mode and frequency.

A weight suspended by a length of rubber cord will bob up and down if pulled down and then released. This motion takes place almost entirely in one mode since vibrations in extraneous modes would require high-frequency distortions within the rubber – distortions which quickly die away. The system possesses one very obvious mode and we can readily experiment with it.

The first thing to notice is that the frequency of bobbing up and down does *not* depend on the amplitude. This, of course, is nothing new because we saw that the vibration of the chain in any one mode retained one characteristic frequency as the vibration subsided (that is, as the amplitude grew smaller). But though we cannot alter the frequency by changing the amplitude, we can alter it by changing the system. An increase in massiveness of the weight which hangs on the cord is found to lower the frequency of bobbing. On the other hand, the frequency can be raised by increasing the stiffness of the cord (by hooking on a second cord to augment the original one, for instance).

These results are of general significance. If the massiveness of a system is increased, then *all* its frequencies are reduced – though some are affected more than others. Equally, an increase in the stiffness of a system will raise all the frequencies – again, some more than others. The 'stiffness' of a piano string is provided by the tension in it so that, when a piano-tuner wishes to raise the pitch of a particular string (that is, when he wishes to raise its frequencies), he tightens it. If, though, the 'stiffness' happens to be proportional to the massiveness (as with the weight of a pendulum bob), an increase of mass makes no difference to the frequency. The act of changing the massiveness or the stiffness is, as we have mentioned, that of changing the system and so we should, in general, *expect* different frequencies (and, of course, different modes).

If sufficient is known about the massiveness and stiffness of a system, it will be possible to calculate its frequencies. This is sometimes a matter of fundamental importance to the engineer. With a structure like an aeroplane which can perform all sorts of contortions, though, it may turn out to be a very substantial calculation indeed.

The question of why one should wish to know the frequencies of a structure, a machine or any other type of system can be left until later. Actually, we shall find that a knowledge is not needed of *all* the innumerable frequencies and interest only centres on those frequencies that fall into some relevant range; often we need consider only the *lowest* frequencies.

If the 'system' is a crystal, such as those in some gramophone pick-ups, then the relevant frequencies are likely to be many thousands of hertz. By contrast, the important frequencies of whole structures and of machines in general are usually much lower, being mostly less than 50 Hz and rarely more than 500.

The lowest frequencies of a system can, in fact, be quite small. A clothes line slung between two posts and having plenty of sag, for instance, may oscillate freely at only one or two hertz. An oscillation of this sort was observed during the autumn of 1959 in the grid system of the Central Electricity Generating Board at the Severn Crossing, shown in fig. 26; the frequency concerned was unusually low, being of the order of 0.125 Hz. The crossing has two large pylons just over 1.5 km apart, supporting transmission cables of 43 mm diameter. It was found that, provided it blew from the right direction, a moderate wind would make the cables sway with low frequency and large amplitude in such a violent fashion that cables normally spaced 8 or 9 m apart actually touched, leaving broken strands and burn marks, as well as short-circuiting the electricity supply. (The probable explanation of this behaviour was eventually found and a cure effected by wrapping the conductors with thin plastic tape, thus altering the geometry of the surface presented to the wind.)

This problem of the Severn Crossing was not directly one of *free* vibration, as the passive system was affected in some way by an external agency – the wind. But, typically, the engineers who tackled it required information about those of the system's natural frequencies which lay in the vicinity of the observed frequency.

There is one final observation that we should, perhaps, make about these frequencies of free vibration. It is in the nature of things that any structure has frequencies of free vibration, the average spacing between which gradually decreases. This seems to be in conflict with the previous comment that the frequencies of the bicycle chain gradually assume an even spacing. Actually, the

Fig. 26. The Severn Crossing showing cables slung between pylons more than 1.5 km apart. The cables were made to oscillate so violently by the wind that they touched. (Courtesy C.E.G.B.)

bicycle chain will serve to illustrate the point quite clearly. When we referred to 'even spacing', the inference was that we were only referring to sideways vibrations in the 'limp' direction of such a chain. But the chain can also twist, it can bend slightly in its 'stiff' plane and its individual links and rollers can bend in complicated ways. All these latter motions would emphasize the stiffness of the metal rather than its massiveness so they would have very high frequencies which would start to fill the gaps between those we discussed previously. The very high frequencies of any system form a steady stream.

2.3 Modes

As the bicycle chain experiments demonstrate, the frequencies of free vibration are closely bound up with the modal shapes of a system, and a vibration analyst instinctively regards them as inseparable. As we have seen, the essential feature of all the characteristic frequencies of a given body is that each of them is associated with a different shape, or 'mode', of vibration. In the example of the piano wire, 440 Hz corresponds to a single loop and there will be other vibrations possible, at 880 Hz with a different shape (two loops), and so on. This is illustrated in fig. 27.

The shape corresponding to the lowest frequency of a pendulum is obvious; the arm remains very nearly straight. The next mode is more complicated since the pendulum forms a loop and oscillates at a much higher natural frequency. There are others corresponding to more loops with higher frequencies still. The first few shapes of a particular sort of 'pendulum' – a uniform hanging chain – are shown in fig. 25: they are the shapes that are obtainable with the bicycle chain.

The piano string and the pendulum are simple systems whose shapes are fairly easy to calculate. Structures like buildings,

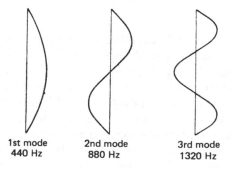

| 1st mode | 2nd mode | 3rd mode |
| 440 Hz | 880 Hz | 1320 Hz |

Fig. 27

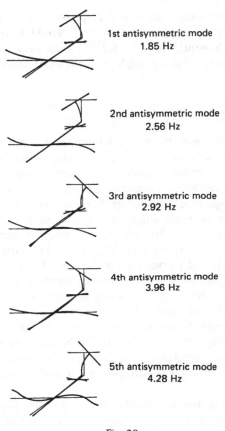

1st antisymmetric mode
1.85 Hz

2nd antisymmetric mode
2.56 Hz

3rd antisymmetric mode
2.92 Hz

4th antisymmetric mode
3.96 Hz

5th antisymmetric mode
4.28 Hz

Fig. 28

machine tools or similarly complicated engineering systems have much more complex shapes. A few modal shapes of the VC 10 airliner are sketched in fig. 28; they correspond to the frequencies we noted earlier. In practice, the modes and frequencies of a new aeroplane are found both by calculation and by experiments.

The modes shown in fig. 28 were calculated for the aeroplane *in flight*. If we seek to find those same modes by means of experiments performed on the ground something has to be done about the way the machine is supported; the aircraft in flight has modes which differ somewhat from those of the same aircraft when it is held up

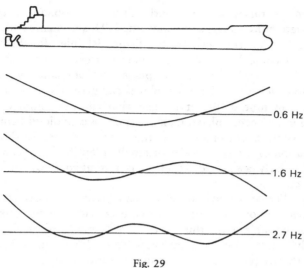

Fig. 29

by its undercarriage. Corrections have to be made. It is the 'in flight' condition that is the more likely to interest the aeronautical engineer. Although it may sound strange, precisely the same problem arises with ships!

Fig. 29 shows a large tanker capable of carrying about 250 000 tonnes of crude oil. Its lowest three symmetric modes of distortion when it is fully loaded but totally unsupported by the sea – as if it were in outer space so to speak – are shown sketched. These modes and frequencies that the ship would possess if it were clear of the water are used in certain types of calculation. But there is no hope whatever of checking these data directly by experiment – only indirectly.

The modal shapes of a system – one for each frequency – have an interesting property. *Any* distortion of which the system is capable can be regarded as the sum of a number of distortions, one in each mode. Thus, if the system is subjected to any static distortion and is then released, it will start to vibrate in all the relevant modes, and each such motion will occur independently at its own frequency. We have already suggested this in connection with the bicycle chain and it explains how the apparently complex free motion of the chain is not incompatible with simple free vibration in modes.

When a system has two or more close frequencies of free

vibration, it raises a particularly difficult problem where the appropriate modal shapes are concerned. This is because the closer the frequencies, the less easily distinguishable do the shapes become. A conical pendulum can vibrate freely in one plane, the '*x*-plane' say, and also in a '*y*-plane' perpendicular to it. The oscillations in each of these two planes can properly be regarded as occurring in modes. But it is clear that the pendulum can also oscillate in any other plane, its motion being capable of being built up from simultaneous motions in the *x*- and *y*-planes. In short, any combination of the two modes originally identified is also a mode. This peculiarly awkward situation is not uncommon in aircraft structures.

The ability of a body to store energy when it is distorted was mentioned as one of the two requirements which make free vibration possible. Now this ability may be modified by changes of the body's state – by a rise in temperature, for instance. This will change the frequencies and, usually, the modal shapes. At super-sonic speeds, aircraft get appreciably hotter as a result of 'kinetic heating' caused by atmospheric friction so that their modes and frequencies change, and this is vitally important to the aircraft designer.

2.4 The decay of free vibration

One other feature of free vibration has been mentioned: namely, the fact that it dies away. The effect is known as 'damping' and it is caused by friction. A bell goes on ringing for a long time after it has been struck because there are no large frictional forces within its material to dissipate energy in the form of heat and because energy is radiated quite slowly in the form of sound waves. On the other hand, if a car is bounced up and down on its springs and is then allowed to oscillate on its own, the movement dies away quickly; the shock absorbers are fitted to ensure that this is so. The springs of a motor-car are suddenly compressed when the wheels run over a bump and, in the absence of the shock absorbers, the body would then commence to bounce up and down unmercifully in free vibration until the energy had been slowly dissipated. As a general rule we expect more violent oscillations in structures which, like the bell, have little damping, than we would in structures whose rate of energy dissipation is high.

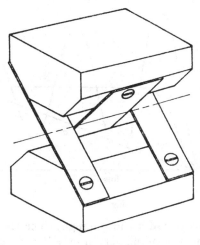

Fig. 30

We shall find that the absence or presence of damping is vitally important in certain systems. Engineers occasionally have to try to eliminate it. It is sometimes necessary, for instance, to use a spring-controlled hinge in a piece of apparatus (such as an instrument) and desirable that this hinge should be as free from friction as is conveniently possible. One way of constructing such a hinge is shown in fig. 30, the axis being shown by a chain-dotted line.

Damping occurs in every vibrating system. It is known for instance that some of the dissipation of energy in a vibrating aeroplane structure is accounted for by the panels of which the plane is constructed; they rub together slightly at riveted joints. A structure like a house is likely to have quite heavy damping, a feature of great interest to those who study the effects of earthquakes.

When damping is a particularly desirable commodity, it may be introduced artificially, as with the vibration dampers on the car. A meter would be little more than a nuisance if its needle swung interminably about the reading that it is supposed to give before settling down to permit the scale to be read (fig. 31(*a*)). Artificial damping is therefore introduced so as to bring the needle to the proper reading without a long delay (fig. 31(*b*)). Too much

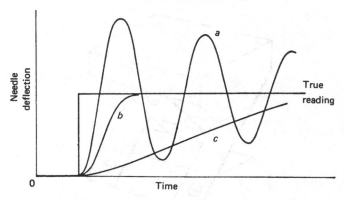

Fig. 31

damping would be as bad as not enough, since the needle would then creep towards its reading very slowly (fig. 31(c)).

Many methods have been used for introducing artificial damping into a system. It may be done electrically, for instance, though there are many purely mechanical devices available. A few of these are:

(1) *Viscous friction in a fluid.* A simple example is provided by a 'dashpot', in which a piston moves in a cylinder and the friction comes from the rush of fluid (often air) through the small gap between the side of the piston and the cylinder wall. In some other arrangements, paddles move in oil or silicone fluid.

(2) *High damping materials.* When a 'bell' made of a certain manganese–copper alloy is struck, it emits a 'thud' rather than a note. Rubber is sometimes used in supports partly for its damping properties. Again fibre blades have been used in gas-turbine compressors on account of their high internal damping.

(3) *Coating on panels.* Preparations are available which may be applied to the surface of a metal panel so that it no longer emits a metallic noise if struck, but only a 'thud'.

(4) *Dry friction, in which surfaces are made to rub together during vibration.* This is used, for instance, in some gas turbine compressors in which the blades are *hinged* to the rotor carrying them. Again, wads of knitted metal wire are placed in some springs to augment the friction.

(5) *Sandwich construction.* Panels made of thin metal sheets separated by a thin layer of viscoelastic material are good sound insulators.

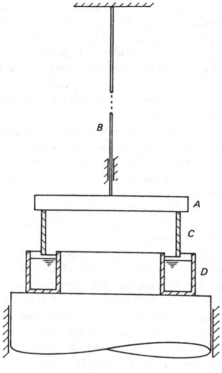

Fig. 32

(6) *Packing with foam plastic and rubber.* An egg or an electric light bulb can be dropped from a great height onto a hard floor without sustaining damage, if it is carefully supported in a suitable packing material.

We may think therefore of two types of damping – that applied deliberately and that which just happens. While it is sometimes possible to make sensible calculations of the damping that is inserted purposely, the damping which occurs otherwise almost invariably defies calculation and needs to be measured.

Having discussed briefly what damping is and how it arises, we must now consider its effects. In particular, how does it affect the frequencies and the modes of free vibration?

Fig. 32 shows a 'torsion pendulum'. It is like the mass on the rubber cord in that it is a system whose lowest mode is clearly identifiable and whose lowest frequency is well separated from its nearest neighbour. A heavy metal disk A is supported by a wire B and so can perform twisting oscillations about its central axis. The disk carries a 'skirt' C which becomes partly immersed in the oil held by an annular container D when the container is offered up to it. If D is held up, the rotation of A becomes much more highly damped than if D is lowered.

Using the torsional pendulum with D lowered out of the way, we may set A into free vibration by twisting and then releasing it. The motion dies away very slowly and we may count out loud as A executes 1, 2, 3, . . . complete cycles. If, while this is going on, D is raised, we discover that, whereas the motion immediately starts to decay far more rapidly, the rate of counting is not perceptibly changed. In other words, the application of quite heavy damping hardly affects the frequency.

To examine the effect of damping on modal shapes, we may use a modified version of the hanging bicycle chain. In the Scotch yoke apparatus of fig. 24, we now hang a second chain from the same yoke a centimetre or two behind the first. Now one chain can be hung in a deep glass tank of paraffin while the other hangs, as before, in air. By looking through the tank we can see two chains – one much more heavily damped than the other.

If the speed of the motor driving the Scotch yoke is varied, a speed is found at which the motions of *both* chains become magnified, though the magnification is greater for the chain in the air. If the motor is now suddenly stopped, the two chains commence their free vibrations. The chains retain their own vibration shapes, which are seen to be the same for both of them. (Both, too, perform the vibration with the same frequency – though we are now prepared for this in the light of the torsional pendulum experiment. And, of course, the heavy damping makes the vibration of the chain in the paraffin subside much the more quickly.) We conclude, then, that the increase of damping neither alters the frequency nor the mode shape of free vibration very much.

These results are typical of damping in general – our experiments are just convenient illustrations. This being so, they suggest an interesting, and very important, line of reasoning. What we have shown is that *increases* of damping make little difference to the modes and frequencies of free vibration. We have *not* shown that the introduction of any damping *at all* makes little difference; even so, it is tempting to assume that such would be the case, and theory predicts that the assumption would be a sound one.

It is convenient therefore, when studying the free vibration of a system, to start by ignoring friction. One is left with an imaginary system whose free vibrations would never decay. But the imaginary system would still possess modes and frequencies – they are called the *principal modes* and *natural frequencies* of the real (damped) system.

Although this is a subtle point, it offers the vibration analyst enormous advantages in practical calculations.* His calculations of principal modes and natural frequencies are not hampered by the complications that damping brings with it. And, when he has found the principal modes and natural frequencies, he knows that they will differ but little from the vibration shapes (or 'modes') and frequencies that he will actually be able to observe when damping *is* present.

We have seen that the modes of a system form a sequence, and that each is essentially different from all the rest. In mathematical jargon, every mode is said to be 'orthogonal' to all the others and the condition of orthogonality can in fact be given mathematical expression. Now in theoretical work this condition is important and it is therefore of profound significance that the condition of orthogonality is very much simpler for the undamped principal modes than it is for those modes which are observable when damping is present.

This idea of imaginary undamped systems can be used to summarize our findings. The amount and distribution of 'massiveness' and the amount and distribution of 'stiffness' in a system between them define the principal modes and natural frequencies of the system. Each principal mode is identified with its own particular natural frequency. Any practical system can be made to oscillate in a mode that approximates closely to the corresponding principal mode and at a frequency that is nearly equal to the appropriate natural frequency, the small discrepancy in each being due to the presence of friction.

* The reader may have noticed that this device also offers a handsome reward in logic, too. It removes the necessity of defining accurately *exactly* what is meant by the 'frequency' and 'mode' of a non-sinusoidal (decaying) oscillation. This is one of the matters that we have carefully avoided; we see now that it is of theoretical, rather than practical interest.

2.5 Free vibration in engineering

Free vibration is a common phenomenon and engineers sometimes make measurements of it – usually in order to obtain information on its modes, frequencies and rates of decay. Such measurements are made, for instance, on some prototype aircraft in flight, the initial disturbance which starts the free vibration being provided by small explosive charges (or 'bonkers'!). Another, more common, method of starting the motion is to administer a jerk to the controls.

In the early days, when life was simpler than it is now, the free vibration of aircraft on the ground was sometimes set up by the sudden release of a steady force. A wing was pulled out of its equilibrium position by means of a rope and, when all was ready, the rope was cut. Observations could then be made of the free motion.

We shall discuss, later on, the swaying oscillations that chimney stacks sometimes perform in the wind. Experimental data have been obtained on the behaviour of a chimney in free oscillation, the method used being to fire small rockets fixed to the top of the stack, as shown in fig. 33, and to study the resulting motion.

Free vibration gradually dies out, sometimes to leave only a steady vibration which arises from some other cause to dominate the scene. When the engine of a car is either started or stopped, the whole engine lurches. This lurching is associated with the actual starting and stopping and does not persist while the engine is running. (The *steady* vibration is something we shall discuss in the next chapter.) The free vibration that is associated with starting and stopping can also be associated with changes of speed. The car engine will lurch when the running speed is suddenly increased, for instance. One of the major causes of free vibration is change of steady operating conditions in machines. It happens, though, that this is not usually of much interest since it quickly dies away as a result of damping.

We shall see in the next chapter that flexible shafts have certain dangerous speeds at which they bow out whilst rotating. When some shafts of this sort are brought up to their service speeds, they have to pass through these dangerous speeds. Just as the change of speed of a car engine sets up a free vibration, so does change of

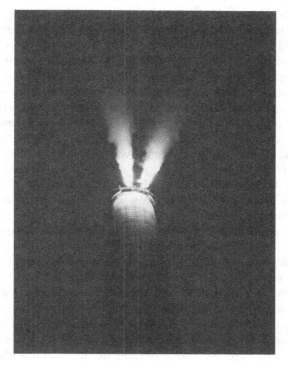

Fig. 33. A chimney being made to sway by the firing of small rockets fixed to the top. By this means it was possible to study the behaviour of the stack in free vibration.

rotational speed cause free vibration in a flexible shaft. The vibration of a large turbo-generator rotor is so serious that the rate of passage through one of these dangerous speeds during the process of running up and of stopping has been closely studied.

These are all examples of ways in which free vibration *as such* may engage the attention of engineers. As we have already mentioned, however, this type of motion has a more profound importance. Its principal modes, natural frequencies and dampings really define the 'dynamical personality' that a system possesses. Thus if we know sufficient about this personality, we have some hope of predicting how the system will behave in various circumstances; if we do not know enough, we have almost none. This is really what much of this book is about.

This possession of a 'personality' can be illustrated in many ways, a dramatic example being provided by the 'roaring drainpipe'. Hold a length of drainpipe vertically and stuff a loose wad of metal gauze up into the lower end. If the gauze is heated for a while with a bunsen burner and the bunsen is then removed, the device will commence to emit a sound. Now the point is that the sound is not just some haphazard *noise*, but has a distinct, characteristic, particular pitch. It is clearly a crude note in the musical sense. The pipe and its enclosed air must therefore have some inherent vibrational properties that determine what note this shall be – some sort of 'personality', in effect. If we wish to investigate the behaviour of the 'roaring drainpipe', we should have to inquire what these properties are, and to ask what its principal modes and natural frequencies are.

This all sounds a bit mysterious. The point is that the 'dynamical personality' of a system largely determines how it will react when stimulated. It is as though mechanical systems would like to oscillate freely in their principal modes and at their natural frequencies. They do not do so normally because they are damped, but they will if given a little enouragement. As we shall see this raises two main possibilities. Either systems can be given this type of encouragement, or their willingness to vibrate freely at a natural frequency can *itself* lead to provision of the stimulus that they crave.

2.6 Some complications

It may well seem, on a first reading, that the foregoing description of free vibration is complicated enough and that things could not possibly be any more complex. But they can – although any features that we have not dealt with so far can, perhaps, be regarded as refinements. Let us briefly glance at these refinements, without going into much detail, just to acquaint ourselves with their nature.

As we have seen, there are three essential properties of systems that are relevant to free vibration. In the broadest possible senses of the words these are mass, stiffness and damping. All three can be modified in a large number of ways. For instance, marine engineers have some difficult problems to face in ship vibration because a fluid flowing past a system can augment its mass, its damping and

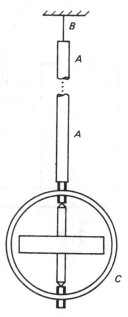

Fig. 34

its stiffness. This is also true of a fluid contained in a system; the fuel sloshing about in its tanks can greatly complicate calculations of a rocket's behaviour in vibration. There is no necessity for us to go into the mathematical details of these problems; the matter is very complicated.

If the vibrating system exhibits gyroscopic effects its free vibration is again out of the ordinary. This may be illustrated by means of a conical pendulum like that shown in fig. 34. The pendulum arm A is supported by a short length of thread B and carries a toy gyrostat C fixed to its lower end. If C is not rotating, the pendulum can be made to oscillate in a single vertical plane through its point of suspension. But if the gyrostat is running, the vertical plane in which the oscillation takes place itself rotates.*

---

* This is most easily seen if a small inverted conical container of dry salt is carried below C. A fine trail will be left by salt crystals running out of a small orifice in the bottom of the container. The trail is most clearly visible if it is made on a black sheet of paper; it is rather pretty.

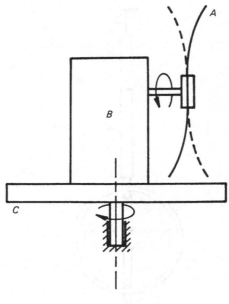

Fig. 35

Some understanding of this behaviour may be had from the apparatus shown in fig. 35. A disk of thick drawing paper *A* (shown in side view) is driven round at high speed by an electric motor *B* standing on a turntable *C*. If *A* and *C* rotate in the directions shown by the arrows the paper bends out at the top and in at the bottom as indicated. If now the direction of *C*'s rotation is reversed, the paper is bent as shown by the dotted line. Thus, if *C* were to perform twisting oscillations about its vertical axis, the distortion of the paper would fluctuate about an axis that is perpendicular to the plane of the page. The effect is thus associated with an axis that is perpendicular to both the axes of rotation; it is this fact which makes it so unexpected, and which makes the gyropendulum rotate as it swings.

The last two complications that we shall refer to both relate to damping. First of all, the effects of *very heavy* damping are such as to destroy some of the simplicity of the foregoing results. It can no longer be relied upon to leave principal modes and natural frequencies almost unchanged and marked changes can occur in both. It is not easy to show this in a demonstration because the friction has to be so heavy that the oscillations decay rapidly. This, in turn, means that observations are difficult to make and that one is forced to think again about what is meant by a 'mode' and a 'frequency' of a far-from-sinusoidal motion.

The second complication that damping can introduce is even worse. So far we have assumed, on the basis of rather crude experimental evidence, that motion in each of the principal modes of a system is made to decay by friction – and does so *independently of all the other principal modes*. But this is not necessarily true and difficulties are sometimes encountered as a consequence.

To take an extreme example of what this can imply, consider the propulsion system of a motor vessel. The twisting principal modes of a ship's diesel drive which embodies a fluid coupling fall into two separate groups. On one side of the coupling is the engine and on the other is a reduction gear, the propeller shaft and the propeller; when all damping is removed for the purpose of defining the principal modes, the connection between the two portions is effectively broken. But if, by some magic, we could distort just one side of the complete system accurately in one of its principal modes and then release that side, the ensuing vibration would become very complicated. The fluid coupling would ensure not only that the other side of the system was thrown into oscillation, but also that the initial distortion would be joined by distortions in other principal modes. The damping is said to 'couple the principal modes'. Such behaviour – and there is certainly no need for there to be a fluid drive for it to be observed – introduces a peculiarly difficult set of circumstances in practical vibration analysis; but, luckily, it seldom matters much.

Imposed vibration

Yes, I like to see a tiger
From the Congo or the Niger,
And especially when lashing of his tail.

It is now possible to distinguish between various types of vibration. The first that we shall discuss is called 'forced vibration' and it is set up in a vibrating system by applying to it a pulsating disturbance of some sort. The essential feature of this excitation is that it remains present and unaltered whether the system to which it is applied vibrates or not.

3.1 Resonance

Let us consider a sinusoidal fluctuation of force with some particular frequency. If it is applied to a system, then that system will vibrate with a sinusoidal motion of the same frequency. This equality of frequency is characteristic of this type of motion.

The shaking motion that we discussed at the end of chapter 1 is a form of forced vibration. The fluctuating buoyancy force acting on the small boat and the force of unbalance in the short rotor also cause vibrations which have the same frequency as the excitation. Why then should shaking be mentioned on its own? The reason is that the force which causes a shaking, causes a perceptible motion when its frequency of application is well below the frequency at

Fig. 36. A simple frame which executes forced vibration when the small electric motor fixed to it rotates. By varying the speed of the motor a succession of modes may be brought to resonance.

which the vibrating structure can oscillate freely, with distortion, on its own.

If the small boat were given a suitable blow with a hammer, then it would commence to vibrate on its own, though the motion would be very heavily damped and would soon die out. The lowest natural frequency of such motion would be far higher than the frequency with which the boat rises and falls bodily on the waves. As to the short shaft of fig. 22– well, we must come back to that later, in section 3.4.

When encouraged to vibrate at a *natural* frequency, systems joyfully respond. If, then, a pulsating excitation is applied with – or almost with – a natural frequency, then a violent motion may be expected. This magnification of the motion is known as 'resonance'. The difference between the forced vibration of shaking that we mentioned previously (in chapter 1) and forced vibration as it relates to resonance is just a matter of the relation between the exciting frequency and the lowest natural frequency.

The structure shown in fig. 36 is a simple arrangement of metal strips joined together to form a sort of model building frame. A small electric motor mounted on it drives a wheel carrying an unbalanced mass. If the speed of the motor is gradually increased, resonances appear successively in the first, the second, the third, ..., modal shapes. This can be shown clearly by throwing a shadow of the frame onto a screen; at each resonance, parts of the shadow become blurred. It is worth noting that this structure is an exceedingly simple one by engineering standards. Even so the calculation of, say, the lowest half dozen of its natural frequencies and principal modes would warrant the use of a large automatic computer; it would be a tremendous task without one.

The frame performs a slight forced vibration as the motor speed increases from zero to the vicinity of the first natural frequency. Thereafter the forced vibration is such that resonance becomes a possibility – that is, the driving frequency may be near a natural frequency. And this is the essential feature of this type of motion.*

In order to achieve near-equality of a natural frequency and a driving frequency, we have adjusted the latter. There is no reason why resonance should not be brought about by adjusting a natural frequency of a system, however, although this must mean making some alteration to the system itself. This can be demonstrated by holding a vibrating tuning fork over the open end of a tall slender tube containing water. The column of air above the water surface is made to vibrate and, by suitably adjusting the water level, one can give the air column a natural frequency equal to the driving frequency – that is, the frequency of the fork. When this is done, the sound of the fork is greatly enhanced. We shall see later how this type of adjustment is sometimes put to use.

Forced vibration can be set up in a very large number of ways. For instance the shaking near the stern of a ship is caused by hydrodynamic forces which are generated as the propeller blades pass through the non-uniform flow caused by the presence of the hull.

A completely different source of excitation occurs in a Pelton wheel. This is somewhat like a very well-constructed water wheel, with accurately shaped buckets attached to its rim; a tangential jet

---

* The word 'shaking' was previously used in the following sense. The boat, the reciprocating engine and the short rotor in section 1.6 were oscillating below their first *non-zero* natural frequency. Their 'rigid body' motion took place, in a perfectly valid sense, in their lowest modes – modes which involve no distortion and which correspond to a zero natural frequency.

Fig. 37. Ground resonance test of an aeroplane. The structure is excited by electrically operated shakers which operate with controllable frequency. Modes are brought to resonance and then measured by pick-ups that are distributed over the structure. (Courtesy Vickers-Armstrong (Aircraft) Ltd.)

of water, usually of very high speed, plays on to the buckets and drives them round. Any one bucket thus receives a periodic blow. If the frequency of impingement of the jet, or the frequency of one of the *components* of this far-from-sinusoidal force, were to coincide with a natural frequency of a bucket, then that bucket might be sufficiently excited ultimately to break. This has, in fact, happened once or twice.

Resonant vibration is one of the causes of blade failure in turbo-machinery, as, in steady motion, any given blade passes any given point at accurately fixed intervals. If, for instance, a water-turbine blade receives a splash, then it will do so at regular intervals and may break as a consequence of resonance.

A forced vibration usually becomes significant only if resonance occurs, and this may be very useful. Fig. 37 shows a VC 10 airliner undergoing a resonance test. Shakers are applied at suitable points

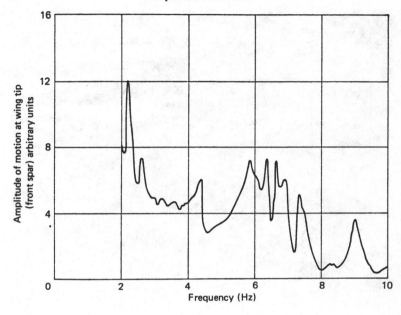

Fig. 38

and the frequency of shaking is gradually varied. When the frequency comes into coincidence with a natural frequency, resonance is set up and the vibration becomes perceptible. The known frequency of the shaker is therefore equal to a natural frequency and by making measurements of the distortion at various points on the structure, the corresponding mode may be found. With the VC 10, tests were carried out over the range 0–100 Hz and some 30 modes were identified in the range 0–25 Hz.

Fig. 38 shows a record obtained from a ground resonance test. Driving frequency is plotted horizontally and the amplitude of motion at some point on the aircraft is plotted vertically. Each peak corresponds to the coincidence of the driving frequency with a natural frequency.

It happens, for various reasons, that the theory and practice of resonance testing are extremely complicated. Reliance cannot always be placed on results which are displayed in the form of fig. 38 and other techniques are used. One reason for this is that the natural frequencies of an aeroplane lie fairly close together and it would be quite impossible, without very special precautions, to excite one mode without perceptibly exciting others. Again, the distribution of damping in an aeroplane is such that motion in one particular mode sets up damping forces which tend to excite other modes. Yet a third difficulty is the very practical one of sheer complexity. Thus the test shown in fig. 37 involved readings from 150 instruments which were deployed around the aircraft structure, and these readings were made with frequency increments of only 5 per cent.

There are many ways, other than resonance testing, in which resonance may be put to good use, most of them far less complicated than this. To shake a sieve mechanically, for instance, it is natural to mount it on springs and to obtain the assistance of a resonant condition.

To sum up then – a fluctuating force will give rise to a vibration which has the same frequency as itself. If that frequency happens to coincide with a natural frequency then resonance will occur and the vibration will be violent. Anything like an exact study of the phenomenon may be a complicated business, particularly if the system concerned has two or more close natural frequencies.

3.2 The suppression of forced vibration

We have seen that resonant forced vibration may be useful on occasion. Often it is decidedly a nuisance. Consider for a moment the frame shown in fig. 36 and imagine that it is erected to support a machine of some sort. If the builder installed a large unbalanced machine on this supporting frame then it could set up resonant vibration, as we have seen. It is clear that, if he knew the natural frequencies of the structure and the shaking frequency of the machine, the builder could see whether or not resonance was likely to occur. The determination of the natural frequencies of frames is therefore important and this is an argument for calculating natural frequencies; but why shapes? The answer is partly that the two calculations usually go hand-in-hand and partly that the extent to which violent oscillation can be elicited at a particular natural frequency depends upon where the machine is placed in the structure relative to the mode shape.

As might perhaps be expected, the heavier the damping in a structure the less violent will a resonant vibration be, for a given excitation. The demonstration model in fig. 36 is an extreme case, and if it were the steel skeleton of a framed structure the brick cladding that would surround it could be expected very materially to increase the damping and therefore to diminish the oscillation.* This fact probably saves many buildings from being well-nigh unbearable.

The damping in some practical structures is fairly light – as, for instance, with suspension bridges. In this event, quite a small pulsating force can produce a dangerous resonant vibration. Thus soldiers marking time are sometimes used as vibration exciters during structural tests of bridges. Great care has to be taken when this is done and, normally, even a small party of soldiers will stop marching and 'break step' when it comes to a bridge. If the cadence of their marching were to correspond to a natural frequency, then it is possible that they might even destroy the bridge. This actually happened in 1831 in Manchester when 60 men broke the Broughton Suspension Bridge over the River Irwell. It also happened in Chatham in 1868 when a party of Royal Marines caused a trestle bridge to collapse. But the biggest tragedy was in 1850 when the Angers Suspension Bridge was destroyed by a French Infantry Battalion, about 500 strong. The collapsing bridge plunged the men into a ravine and 226 were killed.

Two distinct methods have been suggested for diminishing unwanted resonant vibration. The first is to 'detune' the system so that the natural frequencies are shifted away from the exciting frequencies or, as with the soldiers on the bridge, vice versa; the second is to increase the damping artificially. (A third possibility will be referred to later whereby the violent oscillation is purposely transferred to a place where it can do no harm.) Of the two methods the first is usually by far the more effective, though it cannot always be employed since the excitation may have a wide range of frequencies, as with a motor-car engine. The practical calculation of changes in natural frequencies to be expected from alterations of stiffness and mass of a system is therefore vital.

This is hardly the place to take up matters of pure theory, but it is worth mentioning one very pleasing feature of most systems. It happens that the damping has to be very heavy indeed for it to shift the frequencies at which resonance occurs perceptibly away from

* The cladding would also modify the natural frequencies substantially.

the natural frequencies which the system would possess in the complete absence of damping. The damping has to be so heavy that violent oscillation is in any case unlikely, and therefore practical calculation of resonant frequencies is no different from the calculation of natural frequencies.

Where detuning cannot be used to overcome all resonant vibration, dampers may be employed. The best known of these is the Lanchester damper which is attached to the crankshafts of some cars. Attaching a damper to a system does, of course, have the effect of augmenting (and so changing) the vibrating system; this is a topic we shall return to. The Lanchester damper is simply a device which will rotate bodily but which dissipates energy by friction when subjected to a superimposed torsional *vibration*. The point of attachment of a damper is important, for if it were placed at a 'node' – which, like the middle of a concertina, does not take part in the vibration – it would not oscillate (and hence dissipate energy), no matter how violent the vibration of the crankshaft. A knowledge of modal shapes is therefore called for.

3.3 Displacement excitation

If the point of suspension of a pendulum is oscillated backwards and forwards horizontally, the pendulum will start to swing and will take up the frequency of that movement. It thus executes a form of forced vibration – caused by 'displacement excitation'.

Some small object in a motor-car will sometimes oscillate violently when a slight vibration from the engine permeates the whole vehicle. This may occur, as we have already remarked, with the rear-view mirror. The mirror is so small that it is unlikely to affect the dynamics of the whole vehicle appreciably, and the analysis of the motion is quite straightforward; we may assume that the mirror does not affect the vibration of the car *at all*. That is to say we can assume that the motion of the point of support of the mirror is *inexorable*. This assumption is comparable with one that we made previously, namely that force excitation remains present *and unaltered* whether it causes a system to vibrate or not.

The forced vibration to which the mirror is subjected is capable of exhibiting resonance, in which event the small motion of the support causes a large vibration of the rest of the mirror. This resonance is what is so objectionable and it occurs at those frequencies of free vibration that the mirror possesses when its support is fixed.

This assumption of inexorable motion is also made when a vibration-measuring instrument is attached to a structure whose motion is to be recorded. It would be misleading if the measuring device were so large as to affect the motion that it is supposed to measure.

Fig. 39 shows a hand-instrument with a reed whose length may be adjusted so as to vary its ('fixed-end') natural frequencies. Suppose that the length of the reed corresponds to a lowest natural frequency of, say, 15 Hz when the instrument's reed oscillates violently. In this event the frequency of the motion imparted to the point is known to be 15 Hz. The instrument must be much smaller than the vibrating system to which it is applied, however.* Here is a system in which a natural frequency is purposely adjusted to be made equal to a driving frequency.

The valves of a motor-car engine are held shut by compressed springs. When a valve is opened, its springs are compressed a bit more. The springs may be thought of as being subjected to displacement excitation. If this motion becomes resonant, 'surging' takes place in the springs; this has even been known to become violent enough for valves to bounce. Fig. 5(b), it will be remembered, shows a broken valve spring.

The Scotch yoke mechanism sketched in fig. 24 will demonstrate several features of displacement excitation. This is particularly easy when two equal chains are hung from the yoke (as we mentioned in section 2.4), one being damped by immersion in paraffin and the other hanging in air. When we started a free vibration by means of a resonant displacement excitation which was suddenly removed (section 2.1), we were merely making use of the fact that the resonant motion takes place in a mode that approximates closely to a principal mode.

Of course there is no such thing as a truly inexorable motion; to some extent, the small body affects the larger body which causes it to move. In fact displacement excitation can take place when neither of the bodies is small. Suppose, for instance, that a ship's propeller is driven by a steam turbine. The turbine runs at a higher

* This particular instrument works by *resonant* displacement excitation, whereas many vibration devices do not rely on resonance to amplify the measured motion.

Fig. 39. A vibration-measuring instrument with a flexible reed whose effective length is adjustable. The instrument is held onto a vibrating structure and the reed is adjusted until it vibrates violently at resonance. The exposed length of the reed is then a measure of the frequency.

speed than the propeller does and the speed reduction is effected by a pair of gear wheels which mesh in a gear box. If one of the gears were badly made or were offset slightly on its shaft, then the gear wheels would suffer a periodic advancement and retardation as they went round. This in turn would set up a torsional motion of the shafting which would be superimposed on the steady rotation which drives the ship along. This type of motion has been observed and can easily be shown to become violent under the resonant conditions that prevail when the frequency of the relative movement at the mating gears has the same value as a natural frequency of the whole shaft system. It was in a system of this type (though one on land rather than at sea) that the breakage shown in fig. 5(a) occurred.

The cures for unwanted resonant vibration due to displacement excitation are just the same as those for forced excitation; we are merely discussing two ways of looking at the same problem. We can 'detune' by modifying the system, or we can increase the damping. A homely example of the second of these two cures is sometimes said to reduce the spilling of coffee from a cup in a swaying railway train – a form of waste which is clearly the result of displacement excitation. If a spoon is put in the cup, there is a greater loss of energy from the oscillation owing to the formation of eddies. This, at least, is the theory; some carefully conducted laboratory tests have failed to produce much supporting evidence for it.

3.4 Whirling of shafts

The long thin shaft shown in fig. 40 will vibrate freely sideways (rather like a piano wire) at a far lower frequency than the short dumpy rotor shown in fig. 22 would. It is hard to imagine the latter as behaving like a wire, because it is so stout. The long thin shaft is unbalanced since it could not possibly be made perfectly straight nor have its mass axis perfectly coincident with its geometric axis, so that when it is run up to speed it will tend to wobble as the short fat one does (see section 1.6). But now the vibration caused by unbalance can approach and pass the first natural frequency.

When the rotation speed, and thus the frequency of the inertia force, is in near-coincidence with a natural frequency – in this case the lowest – a state of resonance is set up. That is to say the system

Fig. 40. A thin shaft which vibrates laterally when it is driven round at 'critical' speeds.

is stimulated at the frequency with which it would like to vibrate freely. The shaft responds eagerly to this form of coaxing so that, at resonance, it whips violently, exhibiting a phenomenon known as 'whirling'. It behaves like this when the rotation speed coincides with *any* natural frequency. The particular shape into which it bows out depends upon the natural frequency with which the rotation speed happens to coincide. The speeds concerned are called 'critical speeds' and the shapes corresponding to the first few critical speeds are rather like those shown in fig. 27.*

The conventional turbo-alternator, which generates electricity, is a monster whose rotor can weigh more than 70 tonne-f. To

* This behaviour can be demonstrated very clearly at a dozen or more critical speeds if the 'shaft' is a length of spiral curtain wire a couple of metres long.

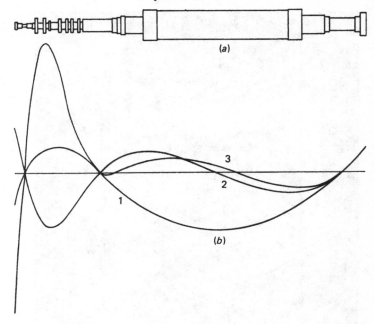

Fig. 41

maintain a 50 Hz supply, a steam turbine has to drive a rotor like that shown in fig. 41 at 3000 revolutions per min. (They rotate at 3600 revolutions per min in the U.S.A. where the mains frequency is 60 Hz.) Fig. 41(a) is a diagram of a fairly large alternator rotor that weighs about 60 tonne-f.* In service it generates 350 megawatts of power at 50 Hz. The calculated shapes of its first three modes are shown in fig. 41(b). The accurate calculation of the dangerous speeds is not easy and yet the rotor must be run through two of them. If this rotor were to get into a sufficiently violent wobble and were to break loose, it would do hundreds of thousands of pounds worth of damage and doubtless account for human lives. One day electricity wlll be generated without the use of these machines; it will be generated directly from thermal energy without a mechanical device having moving parts. This is still a very long way off so far as commercial respectability is concerned, but when it does come, it will remove one of the nastiest and potentially most dangerous of all vibration problems.

* That is, as much as the unladen weight of about eight London buses.

Violent oscillation of an alternator rotor due to whirling must be prevented by judicious balancing. The rotor is safely buried in a heavily reinforced hole in the ground while this sort of work is done (and also while the rotor is driven at over-speed to test its ability to withstand the enormous centrifugal stresses). Fig. 42 shows an alternator rotor in a balancing pit that has its heavy top rolled back. All the instruments and controls are placed in a control room some distance away.

During the process of balancing, small 'heavy alloy' masses are screwed securely into the surface of the shaft, their magnitudes and positions being calculated from measurements of the vibration. It is possible, in theory, to run a shaft up to its first critical speed in order to balance it in the light of measurements made near that speed, and so to pass smoothly through the first critical speed and on up to the second. A suitable balancing technique, which depends on the orthogonality of the modes, then permits the second vibration to be obliterated without upsetting the previously achieved cure for the first. Now the third may be dealt with, and so on. Practical techniques of balancing alternators are still the subject of active research, however, for the problem is not nearly as simple as it may seem. Today, a large alternator rotor on test (unconnected to the steam turbine that will eventually drive it) runs through its first four critical speeds and approaches its fifth. One would expect to have to annul the vibration at the first, second, third and fourth, and also to try to eliminate those that *would* become magnified at the fifth and sixth.

3.5 Vibration of the parts of a structure

The modal shapes and natural frequencies of a system are inherent in that system. If the size of the fly-wheel attached to a shaft is changed, then the frequencies and modes of the whole system of which the fly-wheel is merely a portion are changed. It may therefore seem a little strange that the behaviour of *parts* of systems during vibration are of much concern.

This is not as strange as it may seem. Consider, for instance, an aeroplane whose modes and frequencies are to be found by means of a resonance test conducted on the ground; we have mentioned this earlier. Interest centres on the dynamic properties of the aeroplane in the air, yet the test has to be conducted on the ground

Fig. 42. An alternator rotor supported in bearings in a test pit whose heavy top has been rolled back. The driving motor is hidden behind the far wall. Great care has to be taken when rotors of this sort are run up to, and beyond, their operating speed, for testing. If a large rotor such as this were to oscillate violently at speed it would be very dangerous. (Courtesy C. A. Parsons and Co. Ltd.)

and therefore the aircraft must be supported in some way. Clearly the nature of the supports must not be allowed to influence the results materially. It is found that this requirement is fulfilled if the aircraft is supported on 'soft' springs – a common technique is to let some of the air out of the tyres for the duration of the test. The aircraft without visible means of support can be regarded as a *part* of the system comprising the aircraft and the means of support that are used for the ground resonance test.

It is sometimes possible to limit an excessive resonant vibration by means of a device called a 'vibration absorber'. Suppose that a machine has a particularly violent and objectionable vibration at some particular frequency and that it is specifically required to suppress this motion. It may be possible to shift the dangerous frequency by altering the system in some way so as to change its natural frequency. Alternatively it may be possible to add some device to the system which will take up the excessive motion in a way that does not matter. The original system is to be regarded as part of the augmented system.

The idea behind the vibration absorber may be illustrated in quite a dramatic fashion by means of the apparatus shown in fig. 43. The motor *A* drives an unbalanced arm *B* which imparts a vibration to the member *C*. (The motion of this member causes the arm attached to it to deform the two springs.) The thin 'stalk' *D*, carrying the small weight fixed near its upper end, may be inserted into a small hole in *C* or it may be withdrawn. *D* is the 'absorber'. When it is removed, the rocking of *C* is very noticeable. But when *D* is inserted – as it is in the figure – *C* remains almost stationary while the stalk executes a violent oscillation. Now the motor *A* can only be run at one speed and the position of the weight on *D* is adjusted so that the absorber works at the appropriate forcing frequency. Such a 'tuned absorber', then, is useful at only a single frequency.

Vibration absorbers are, perhaps, not as commonly used as they were. They are sometimes attached to crank webs in internal-combustion engines and an interesting example was to be found in certain of the radial aircraft engines used in World War II. These had a pair of large steel balls that could run in circular grooves, mounted on the crank web. A large motion of a ball in its groove did not matter, and it had the very desirable effect of smoothing out the rotational motion of the crankshaft (which carried the groove).

Fig. 43. The member C of this demonstration apparatus may, or may not, be caused to oscillate violently when the unbalanced arm B is driven round by the constant-speed motor A. It does so if the 'absorber' D is removed, but does not do so when D is attached. If D is attached, D vibrates violently while C remains almost motionless.

Vibration absorbers are used in some washing machines to prevent vibration of the whole appliance, and they have been used in electric hair-clippers to prevent undue transmission of vibration to the hairdresser's hands.

Fig. 44 shows some electricity transmission lines. Small dumb-bell-shaped attachments can be seen, placed on each line about 1.5 m away from the supporting insulators. The attachments are called 'Stockbridge dampers' and their purpose is to augment the damping of the lines in much the same way as the Lanchester damper which is fixed to some crankshafts. As we shall see, the wind can cause the lines to execute small vibrations of fairly high frequency; this sort of motion is quite distinct from that of the Severn Crossing (fig. 26), and it would be far more common but for the widespread use of the dampers. These devices quell the lines' vibration near their ends so that they do not snap at the points of high stress at the insulated supports. Forced oscillation of these cheap little dampers does not matter. Here, then, is another case in which a system is augmented in order to modify its behaviour.

We have begun to see that natural frequencies and modal shapes are of great technical importance. Let us, therefore, return to them briefly, remembering that they are calculated under the assumption

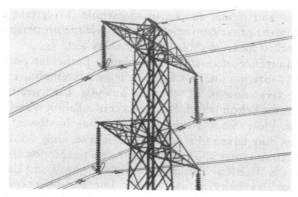

Fig. 44. Dumb-bell-shaped Stockbridge dampers attached to the transmission lines near their insulators. If these dampers are not attached, the wind may cause sufficient vibration of the lines for fatigue damage to occur at the highly stressed points of support. (Courtesy B.I.C.C.).

that the damping is removed by some means. When an undamped system vibrates freely, any given *part* of it executes a forced motion although no stimulation is applied to the whole system. The excitation of a part of the system is supplied by the portion from which it is thought of as being detached. In other words, the considerations which arise in the 'taking to pieces' of freely vibrating systems are very much bound up with the response of the partial systems in forced vibration. The original technical problem from which this aspect of vibration theory originally sprang will serve as an excellent illustration of what is involved.

A piston aero-engine runs efficiently only at fairly high speeds, whereas the propeller which it must drive is efficient only at relatively low speeds. A reduction gear is therefore placed between them. The system of aero-engine, reduction gear and propeller is capable of torsional vibration, and this was a factor to be reckoned with in the days of piston-engine propulsion. The natural frequencies of these systems had to be calculated and of course the calculations could not be made until the systems were adequately known. By imagining the system to be divided up into a propeller on the one hand and an engine with its reduction gear on the other, it became possible for the manufacturer of each portion to supply data which, separately, related only to the forced vibration of the parts. When used suitably this information could give the natural frequencies and modal shapes of the whole. In certain circumstances it might prove convenient to calculate the properties of one portion and to measure those of another by test.

The importance of this idea of isolating particular portions of systems in vibration analysis can hardly be overemphasized, for it lies at the very roots of all vibration analysis. Let us return for a moment to the Pelton wheel. Each bucket of a Pelton wheel receives a periodic blow from a water jet, so that its lowest natural frequencies may have to be determined. It is securely bolted to the wheel which supports it. There is therefore a temptation to say that the bucket is 'fixed' at the junction between itself and the wheel, and to use this fixity as part of the information necessary for calculating the natural frequencies. We already know enough, however, to realize that, at some frequencies, the wheel itself will be very far from rigid and can oscillate freely by itself.

Before any calculations can be performed for a system, it is essential to define the system accurately and also the conditions that will be supposed to prevail at the boundary between the system and its environment. While this problem of the boundary is simple for a body vibrating in outer space, it may cause real difficulties with bodies which vibrate anywhere else – even if they are suspended in the air. For this reason alone, engineers must think not only about forced vibration of whole systems but also of *parts* of systems.

3.6 General periodic forcing

In this chapter, we have usually implied that the fluctuating disturbance – the 'excitation' – varies sinusoidally. While this is commonly a close approximation to reality, it will not always suffice. The excitation may be some regularly repeated but non-sinusoidal disturbance (as with the Pelton wheel bucket) or it may not be regularly repeated at all. We shall consider the first of these possibilities here, reserving the second until later.

Before examining the problem of a regularly repeated (that is, a periodic) disturbance which is non-sinusoidal, we should perhaps notice one more way in which it can arise. This is so common that its importance needs no emphasis.

Fig. 45 shows a simple engine mechanism. The crank A can rotate as shown. It is hinged to the connecting rod B which, in its

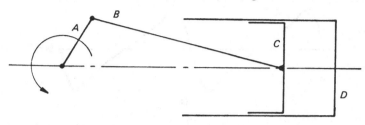

Fig. 45

turn, is hinged to the piston C. This piston is capable of sliding to-and-fro in the cylinder D. Now it is clear that a constant speed of crank rotation will be associated with a fixed frequency of the piston's oscillation. In other words, the piston motion will certainly be *periodic*. It will not be sinusoidal though. As a matter of fact, the displacement graph of C only approaches a sine waveform if the ratio

$$\frac{\text{length of } B}{\text{length of } A}$$

is made indefinitely large. It follows that any force that is associated with the position of the piston – such as the force of inertia, or the force exerted by the piston on the 'gudgeon pin' that holds it on the connecting rod – will be periodic, but not sinusoidal. Many other fluctuating forces of this type are met with in everyday life, as we shall see in chapter 6.

These more general periodic forms of excitation introduce no fundamentally new idea. All they do is bring in complications. It was shown in section 1.3 that a regularly repeated waveform may be thought of as the sum of a number of sinusoidal waveforms, each component having its own amplitude and frequency. This is true of the periodic excitation that we are now discussing; it can be regarded as being several sinusoidal excitations, all acting at once. Any of the components can set up a resonant condition of the type we have already discussed.

This possibility of what might be termed 'selective resonance' has an interesting consequence. It means that the waveform of the response of a system may be quite different from the waveform of the excitation causing it. Consider, for instance, the curve of fig. 46(*a*) and suppose that it represents a fluctuating force. The rather

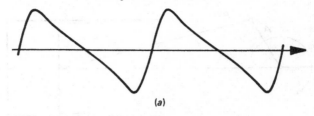

(a)

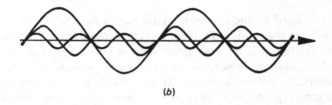

(b)

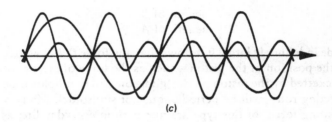

(c)

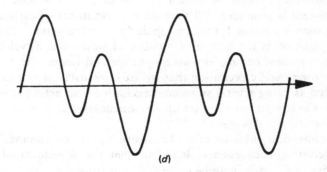

(d)

Fig. 46

Fig. 47. A length of chain C whose topmost link can be thrown from side to side by electromagnets D, the frequency being controlled by the variable-speed motor A and cam B. A harmonic of the excitation can produce resonance in a mode of the chain.

lop-sided waveform of this force is produced by just three sinusoidal components as shown in fig. 46(b). Each of the force components will set up its own forced vibration and the three responses could have the amplitudes shown in fig. 46(c). It will be seen that the excitation component of intermediate frequency is rather near the resonant condition. But if we now add the components shown in fig. 46(c) together we get the total (periodic) response shown in fig. 47(d). The waveform of this response bears little resemblance to that of the excitation that causes it.

The situation is even more complicated than this because, in theory, any one component of the excitation can bring any mode of a system to resonance. (In fact there is no reason why two or more components should not cause resonance in different modes.) With care, this can be demonstrated with the apparatus shown in fig. 47. The electric motor A drives a cam B that actuates a switch. Each time the switch works, the top of the bicycle chain C is thrown sharply across between two electromagnets, D, first one way and then the other. The top of the chain is given a displacement excitation of approximately 'square' waveform in this way, the frequency being adjustable through variation of the motor speed. This excitation has components with 3, 5, ... times the basic frequency as we saw in fig. 11. If the motor speed is adjusted slowly it is possible to cause the chain recognizably to take up, for example, the vibration form of fig. 25(b) at three times the frequency of switch operation. In other words, that component of the excitation which has three times the frequency of the square wave of excitation can be made to bring the second mode of the chain to resonance.

3.7 Random vibration

If the pressure is measured at some point in a turbulent flow of gas – in a heat exchanger, for instance – the value will be found to fluctuate with time in a haphazard manner, as indicated by the curve in fig. 48(a). Such a curve might also represent the surface of a rough road. The pressure fluctuations can cause structural vibration in, say, a pipe, while the roughness of roads is one source of displacement excitation in motor-cars. In analysing motions of this sort – which are very common – it is usual to assume that the vibration is forced; that is, the existence of the excitation does not depend on whether or not it causes vibration.

Before we discuss the response to this sort of excitation, we must think about the excitation itself. Curves like that of fig. 48(a) are usually extremely difficult to obtain accurately, and they are very complicated. The complication of the excitation raises the question

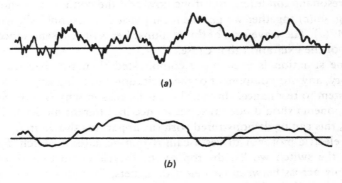

(a)

(b)

Fig. 48

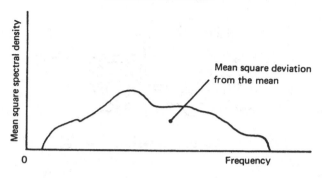

Fig. 49

of how an engineer can make sensible attempts to minimize the vibrations that they generate. In fact he uses an averaging technique.

The first, and obvious, average that can be taken is the 'mean value'; that is, the average height of the curve. But although this gives *some* information it is unable, for instance, to distinguish between a steady (mean) value at one extreme and a violent fluctuation about the mean at the other. This drawback is overcome by the 'mean square deviation from the mean'. We find the mean height of the curve and, in effect, subtract it from the measured fluctuation – leaving a curve of deviation from the mean. If this is now squared (i.e. all deviations from the mean are squared) it gives this second useful average. For shortness, we shall refer to it loosely as the 'mean square'.

The curves of fig. 48(*a*) and (*b*) could well have equal mean values and mean squares, even though they are very different as regards the extents to which they are 'spread out' in time. Now, although it is not obvious that 'frequency' has any relevance in this matter, it does in fact come to our rescue. The mean square can be thought of as being built up of components having *all* frequencies in some range. That is to say, a curve like that of fig. 49 can be plotted, the total area under which equals the mean square; it is a curve of 'mean square spectral density'. The difference between the curves of figs. 48(*a*) and (*b*) would be reflected in the mean square spectral density, that of the latter being more concentrated towards the lower frequencies.

Imposed vibration

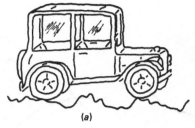

(a)

Periodic time of free vibration of the car

(b)

Fig. 50

An excitation can conveniently be specified in terms of a mean value and a curve of mean square spectral density. Although this admittedly represents only a tiny fraction of all the information that might be given, in practice even this presents severe difficulties of measurement. (To mention but one thing, the length of the curve over which the averages should be reckoned must be 'very long' and a decision is needed as to *how* long will be sufficiently accurate.)

Not only can the excitation be specified in this rudimentary form, but also the response. The system to which the excitation is applied will react in a way that is determined – as we should expect – by its natural frequencies, principal modes and levels of damping. The lower the damping, the greater will be the system's response. In fact, the system seeks out and magnifies the effects of those contributions to the excitation's mean square spectral density whose frequencies lie in the region of its own natural frequency. The response is somewhat like that indicated in fig. 50. The vibrating system 'filters out' much of the excitation and accepts those frequencies that it likes best. This will be reflected in the response mean square spectral density by a pronounced contribution in the vicinity of the system's natural frequency.

This is easy to demonstrate in a rough and ready way by blowing across one end of a tube whose other end is closed. The air in the

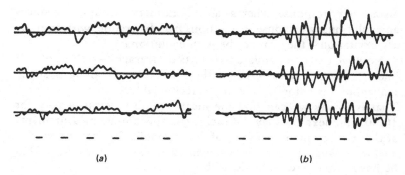

(a)   (b)

Fig. 51

tube oscillates and the result is that a crude, but identifiable note is heard. Indeed if the plug which closes the tube contains a microphone it can readily be shown that the air oscillates in a number of resonant modes simultaneously.

As well as evading many points of theory, we have made one gross oversimplification. It arises from the fact that excitation curves of the types shown in fig. 48 are never *exactly* repeated. This is certainly understandable in the pressure-measuring experiment, as *complete* control of a flowing fluid is not possible. And, so far as the road surface is concerned, no two stretches of road could be identical, right down to the last detail. For this reason engineers turn to statistical techniques – they have to think of *probable* levels of vibration. The simple line of reasoning that we have adopted so far is only valid if the excitation can be relied upon as being completely 'typical' – or 'ergodic'.

For a given excitation we have now to think of an infinite family (or 'ensemble') of all the possible curves, rather than the one 'realization' that we considered before (fig. 48(a) or (b)). Under conditions of steady running, the pressure measurement of fig. 48(a) might be a member of an ensemble like that of fig. 51(a); any uncertainties in the value of pressure are more-or-less uniform, or 'stationary'. On the other hand, if we contemplate excitation whose properties vary, we have 'non-stationary' (or, better, 'evolutionary') conditions as shown by the ensemble of fig. 51(b); this will be the case, for instance, if we conduct our pressure-measuring experiment whilst the operating conditions of our heat exchanger are changing – as during start-up. A stationary excitation produces a stationary response, whereas an evolutionary excitation produces an evolutionary response. Conditions *may* be ergodic with stationary excitation – they cannot be with evolutionary.

Now we can contemplate two sorts of averages – of 'mean' and 'mean square' values. One of these is reckoned along a single realization (as before); the other is reckoned 'across the ensemble'. That is to say we can find the mean square spectral density for 10.30 a.m. next Tuesday, averaging the appropriate measurements at that time from all members of the ensemble. This second type of average is what the engineer normally wants, while the first is what he has normally to make do with.

The reader will no doubt have sensed by now that we are on the edge of an enormous, and mathematically intricate subject. It should be added that all the topics noted in section 1.4 and 1.5 bring fresh difficulties to 'random vibration' – the name given to the type of motion now under discussion. For it is not yet adequately clear what laws govern the failure of metals, for instance, under conditions of random loading.

There can be no doubt that random vibration will be studied very carefully in future. It can be a nuisance, as when shuddering of the needle prevents accurate readings from being made with a fine chemical balance, or when the background hiss of an old gramophone record provides an audible random pressure fluctuation.

The tungsten filaments of electric light bulbs present a practical problem of random vibration. What with being swtiched on and off and with having to carry sufficient current to give us the light we require, these fragile filaments have a great deal asked of them. It is not surprising, then, that their lives may be very much shortened if they are used in a vibrating environment; this is the case when they are placed near the engines in ships, or in motor-car and bicycle lamps, or in aircraft. Designing the filaments of incandescent lamps therefore involves much development work and sometimes calls for environmental testing under service conditions.

Unfortunately, random vibration is also a potential menace. Consider, for example, the noise from a jet engine. Most of it lies in the frequency range 100–1000 Hz and it has an ill-defined peak somewhere in between these limits. The fatigue cracking shown in fig. 52 occurred in the lower surface of the starboard elevator of an aeroplane. The plane had been used to carry a rocket motor which

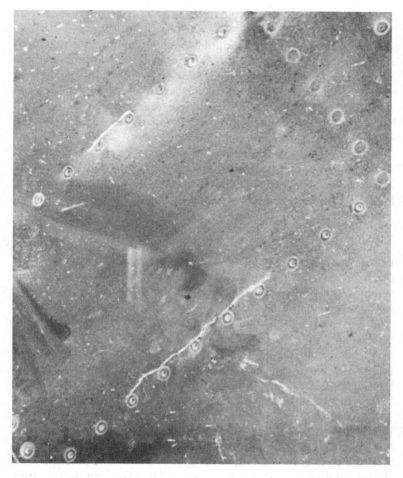

Fig. 52. Fatigue cracking in the elevator of an aeroplane. The damage was caused by noise from a rocket which was directed well away from the tail unit. The fatigue was caused by pressure fluctuations in the sound waves.

was undergoing development flight trials. The motor was attached centrally beneath the fuselage and was directed backwards and downwards – well clear of the elevators. The rocket efflux was intensely noisy and the roar was sufficient to cause the damage. If this sort of thing should be allowed to happen to the pressurized cabin of an airliner – and, of course, it isn't – the results would be disastrous.

If an earthquake tremor should be violent enough to bring down a large building or rupture a dam, the results could be appalling. Yet earthquakes are not predictable or well-defined movements, so that difficult problems arise in the design of large structures in seismologically active countries. Some recent research has been based on the assumption that earthquakes are adequately represented as evolutionary random disturbances.

When it is recalled that every little component of a piece of electrical equipment housed in an aeroplane or a rocket, however inaccessible it may be, is liable to fatigue, it will be seen what an intricate business random vibration is.

3.8 Ships

Let us now briefly think about the motion of something rather particular, namely a ship in waves. By so doing we can perhaps gain some insight into the nature of engineering research; for although we shall be thinking only about one very highly specialized engineering topic, it has all the ingredients that give research work the fascination it has for some.

A ship is a more or less elastic structure and, so far as fluctuating effects are concerned, the sea has two types of influence. First of all, the sea excites the hull, causing it to move bodily and to distort. Notice that here we have an instance in which the excitation passes along the hull and does not remain stationary relative to it. Now the pattern of waves at any instant and the fluctuation of wave height at any given point as time goes on are both of a random nature and so the ship suffers random excitation. A description of the excitation 'input' must therefore be based on information contained in wave atlases and on long-term wave statistics collected over the years by oceanographers. The input is also dependent on the shape of the hull below the water line and on the way the vessel is being operated – on its speed and its heading relative to the waves.

The second type of effect that the sea has on a ship is far more subtle. When the hull moves relative to the water it naturally experiences external forces which would not be applied to it were it not doing so. But the internal forces of inertia, damping and stiffness are of this sort too and it is they which determine the dynamical character of the hull in the absence of the water, as if the vessel were resident in outer space. In other words the sea effectively modifies the inertia, damping and stiffness forces of the dry hull. Unfortunately this does not just mean that the principal modes, the natural frequencies and the damping of the hull suffer quantitative changes as a result of the sea's presence – they are also changed in a qualitative sense. A ship is an example of a 'non-conservative' system.

In all its generality the study of non-conservative systems is a complicated business. But the engineer is chiefly concerned with only two of their characteristics, namely behaviour at resonance and instability. The ship provides an excellent example of resonance, while instability is something we shall take up in the next chapter.

It is perfectly possible, as we shall see later, to administer such a severe blow to a ship when it is under way that it is left shuddering in the sea. The shuddering subsides and we conclude that a free vibration of some sort is performed. It is no longer possible, however, to define *principal* modes and *natural* frequencies as we did previously. The ship merely oscillates in a complicated way with more-or-less identifiable modes, frequencies and rates of decay.

What is rather more to the point here is the behaviour of a ship when it suffers sinusoidal excitation. If a ship proceeds in sinusoidal waves it may be brought to resonance. The wave-encounter frequencies at which this occurs are those of the free vibration and the modes also correspond. Although it cannot be said that the motions can be predicted with as much confidence as one would wish, it is possible to estimate the statistics of a ship's motions and distortions, given the statistics of the seaway, the operating conditions and the appropriate information on the ship.

It would be misleading to give the impression that all this is straightforward and it is worthwhile to dwell briefly on the reasons why this is so. Some of the difficulties are, to be sure, just what one would expect – those of obtaining reliable data on the ship and the seaway and of estimating the inertial and damping effects of the sea (the 'added mass' and the 'fluid damping'). But there are other less obvious matters. How, for instance, can one obtain a measure of the hull damping without having its effects muddled up with the

Fig. 53. A ship that is driven too hard into heavy seas, like this frigate undergoing trials at high speed, may suffer severe deck-wetting. (Courtesy Ministry of Defence.)

rather awkward damping effects of the sea?

A second difficulty is of special importance. The *dry* hull would have six types of rigid body motion with zero natural frequency; they are those of 'surge', 'drift', 'heave', 'roll', 'pitch' and 'yaw'. But when the ship floats in the sea three of those motions (heave, roll and pitch) lose their identity. They cannot be performed without some accompanying distortion and they are no longer associated with zero natural frequencies. Free motions that are predominantly in heave and pitch are usually of nearly equal frequency, are closely coupled and are heavily damped. Indeed such motions would be far from sinusoidal and would raise questions as to what is meant by their 'frequencies'.

Even if all the technical problems could be solved, there would still remain one awkward source of difficulty. A ship will have a succession of masters during its working life and those officers will make different demands on her. To exaggerate a little, what one captain will think are near hurricane conditions requiring great reductions of speed another might think a nice healthy breeze which gives his vessel a jaunty lilt. In other words, even if the sums

Fig. 54. The motions of a ship in waves may make conditions difficult for the crew, affecting their efficiency if not their health. Conditions in this trawler were probably very unpleasant. (Courtesy Ministry of Defence.)

could be done accurately there will remain the question of whether an answer is 'good' or 'bad'.

Even though this type of work is so beset by problems, it has to be done because the designer cannot afford to produce an expensive vessel which turns out to be vulnerable at sea. He needs assurance that his proposed ship will not readily dig its bows into waves, that its motions will not be so violent as to make conditions on board intolerable or that the stress level will be excessive so that cracking of the hull may occur. These various possibilities are suggested in figs. 53, 54 and 3 respectively.

When a ship is actually lost in the open sea this may mean that she has failed to survive the forced random oscillation imposed by waves, either by flooding or by breaking up. If a ship is lost by flooding the clear implication is that she has turned sufficiently far over, either about an athwartships axis (pitching) or about a fore-and-aft axis (rolling) to have presented an opening to the sea. In January 1970, the British lifeboat *The Duchess of Kent* turned end-over-end in short steep waves in the North Sea with the loss of

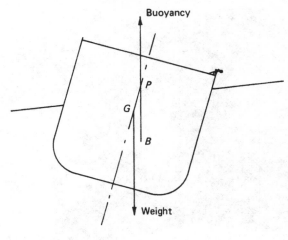

Fig. 55

five lives. But, such are the frequencies normally met with in a heavy sea and normally inherent in ships, rolling is more likely to be magnified than pitching. Many more ships have been lost by rolling too far than by pitching head-over-heels. To quote another extreme case, an American destroyer, the U.S.S. *Hull*, sank in a typhoon in 1944 by flooding *down her stacks*.

The mechanism of rolling is complicated, not merely because the excitation is random but also because the oscillations are 'large' and therefore fall into the category that we shall discuss in chapter 6. A rough and ready discussion is quite simple however. Suppose that the condition at some instant is as shown in fig. 55. The ship has rolled in the clockwise direction and the surface of the sea is inclined to the horizontal. The motion of the ship is governed largely by the buoyancy force and weight.

The weight may be thought of as acting through a point *G* that is fixed in the ship on its plane of symmetry and (usually) near the level of the still-water line. If we assume that the buoyancy force is determined entirely by hydrostatics (and this is unlikely to be far wrong) then we find that it acts through the point *B*. The sea is plainly capable of supporting sea water in the indentation of the surface made by the hull; in other words the surrounding sea exerts an upward force that acts through the geometric centre B of the displaced water – a point which moves relative to the ship.

So long as P is above G in fig. 55 the ship will tend to regain its upright position and it has the 'stiffness' that we have seen is necessary for vibration to be possible in the normal sense. It seems evident, then, that all the designer has to do is make sure that P is always above G.

Unfortunately this is not quite as easy a requirement to meet as it seems. In the first place GP in fig. 55 must not be made too large or the ship will be unbearable to live in since she will be *too* stiff and will roll with what seems a viciously high frequency. It is therefore unfortunate that G may move from its intended position in service, and do so in just the wrong sense. It will be moved upwards by the formation of ice on the upper-works of the ship, for instance, as with a trawler working in polar latitudes. Alternatively it can move to some extent in the athwartships direction if the ship is carrying liquid with a free surface (like fuel, fresh water or even dry grain in tanks). Indeed the possibility of flooding the necessarily large open deck of, say, a car ferry is a very serious consideration indeed because it could easily shift G towards the line of action of the buoyancy force (fig. 55) and so nullify the stiffness of the ship in roll.

It is possible, though fortunately rare, for a ship actually to break up in a heavy sea, like the vessel shown in fig. 3. So far as any normal ship is concerned, only the lowest modes are likely to be much excited by waves because the hull is usually a high-frequency structure placed in a low-frequency environment. A long massive ship, like a loaded supertanker, or a container ship with its large deck opening is the more vulnerable since it has lower resonance frequencies than shorter vessels like tugs, warships and coastal steamers. The responses in the various modes determine the stresses and if those responses are resonant the stresses are high. Notice that the locations of the most highly stressed parts of the hull differ from mode to mode. It is as if Mother Nature is quite determined that we shall never be able to estimate the hull stresses set up in rough sea with any accuracy. Unfortunately she is particularly unkind in this because she has also decreed that sea water shall be corrosive and we have some evidence that repeated stressing in a corrosive medium can be particularly risky.

Quite apart from the danger to the crew of a ship that breaks up, the breakage of a loaded supertanker is an ecological disaster of breath-taking proportions. It would therefore be foolish to ignore the fact that ship strength calculations have hitherto been based on rules with rather little claim to scientific justification.

4

Vibrations that cause themselves to grow

Your badinage so airy,
Your manner arbitrary,
Are out of place
When face to face
With an influential fairy.

The next form of vibration is called 'self-excitation' and it can arise in non-conservative systems. It differs from forced vibration in that if there is no vibration there is no excitation.

We may look at the matter in two ways. The external forces applied to the system are now determined by the motion of the system. But the internal forces governing free vibration are also of this sort, so it is now as if the external forces augment the inertia-, the damping- and the stiffness-forces. The free vibration behaviour is thus modified as if the mass, the damping and the stiffness are changed – as we saw with the ship.

Alternatively we may note that self-excitation requires some source of energy. The vibration is now such that it extracts energy from the source. What decides whether or not a system can tap an energy source in this way? –the system's principal modes, natural frequencies and dampings, that is, its 'dynamical personality'. The very generality of the requirement that only a source of energy is needed to make self-excitation a possibility makes this form of oscillation important and, often, perplexing. Thus each problem of self-excitation is raised by a physical process, the nature of which may not be at all familiar.

4.1 A simple example of self-excitation

Many examples of self-excitation arise from the flow of fluids, the necessary energy input to the vibrating system being taken from the flowing stream. The system shown in fig. 56 demonstrates this. The fan blows a current of air past a wooden bar that is mounted on springs. The bar bounces up and down, perpendicular to the air stream, gradually becoming quite violent.

An explanation of this motion requires a detailed study of the flow of the air around the wooden bar, and here it will be noticed that the bar has a semi-circular cross-section, the flat face being presented to the oncoming air. Suppose that by some means the bar acquires a vertical upward velocity, *no matter how small*. The oncoming air would appear, to an observant fairy sitting on the bar, to have a motion directed slightly downwards. The fuid flow would look somewhat like that depicted in fig. 57. The pressure in the turbulent region shown in the bottom right-hand corner of the diagram is more or less atmospheric, whereas the air in the region directly above the bar is speeded up and can be shown to have a

Fig. 56. A wooden bar of semi-circular section that is suspended on springs with its flat surface presented to the fan. When the fan is switched on, the bar commences to bounce up and down and this motion gradually becomes more violent.

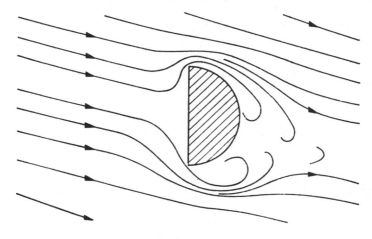

Fig. 57

pressure which is slightly *less* than atmospheric. It follows then that there is an upward force on the bar.

A somewhat similar state of affairs arises in the simple demonstration shown in fig. 58. If air is passed through an ordinary funnel in the direction shown by the arrow, a table tennis ball is held upwards and does not fall under the influence of gravity. This is because the air is speeded up in the circular region where the ball lies closest to the surface of the funnel so that there is a reduction of pressure above the ball.

To return to the semi-circular bar, we see that the air exerts a force on it in the same direction as its velocity, and were the velocity to reverse, then the force would also reverse. The aerodynamic force in this case effectively nullifies the damping of the mechanical system. If the work done per cycle by this force exceeds the energy dissipated per cycle as friction within the system, then the motion will go on increasing until it is limited by the intervention of some extraneous influence (such as breakage!). We see, too, that this will happen no matter how small the initial velocity of the bar. The system is said to be 'dynamically unstable' or, less precisely, simply 'unstable'.

It should, perhaps, be added that this explanation of why the bar bounces is much simplified. Like most problems of fluid dyamics, this one is really more complicated than it might appear. But

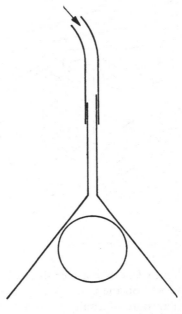

Fig. 58

engineers have quickly to learn that frankly approximate theories are usually far more useful than more complicated ones which have greater claim to 'exactness'. We are told, of Pitti-Sing in *The Mikado*, that

> Her taste exact
> For faultless fact
> Amounts to a disease.

The possibility of contracting this affliction is actually an occupational hazard of engineering.

Although the semi-circular bar is just a toy, it serves to introduce a very real problem, namely that of the 'galloping transmission line'. In certain weather conditions electric transmission lines slung between pylons have been observed to oscillate with great amplitudes and at very low frequency. Although this is not common, it has been observed in the north of North America where ice can form on these lines during hard winters so that the cross-section presented to the oncoming wind is not circular but has a shape that gives rise to this self-excitation. This galloping motion is rather like that of the Severn Crossing that we mentioned in chapter 2, but there is really little connection between the two motions; no deposit had formed on the conductors that had been slung across the Severn.

As we should expect from an adjustment of damping, the bouncing of the bar of fig. 56 occurs at the corresponding natural frequency. Now the bar has several principal modes of low-frequency oscillation, only one of which is the motion that actually takes place. The bar could oscillate backwards and forwards in the direction of the airstream, or move along its own axis or rotate about its mid-point so that one end goes up while the other goes down, and so on. Thus an investigation, starting from scratch, into the possibility of self-excitation would have to be such that the system could select whichever of its possible shapes of oscillation it chose to adopt – including motion in any combination of the principal modes. It is hardly surprising that fresh types of self-excitation are never predicted with accuracy.

4.2 Coupled flutter

There are many ways in which self-excitation can be brought about, and some forms of this motion are essentially more complicated than the one we discussed in the last section. This is because it is not always – or indeed usually – sufficient to regard self-excitation as simply the nullification of damping. In effect, mass and stiffness changes can also occur. But these changes do *not* occur in a way that could actually be obtained by an adjustment of the system's mass or stiffness, as we noted with the ship. The effectively modified mass and stiffness still define 'natural frequencies' and 'principal modes', of a sort, but these do not usually conform to the simple pattern that emerged in chapter 2.

The motion of the semi-circular bar of fig. 56 is said to occur 'in one degree of freedom'. This means that, for instance, the displacement of the centre of the bar from its mean position gives a complete description of that distortion which the spring-bar system chooses to exhibit. Only one *form* of motion occurs and one quantity is sufficient to describe the amount of distortion in that form. Let us now examine a type of self-excitation which demands, for its very existence, motion in more than one degree of freedom.

Fig. 59. A simple aerofoil. This system can deform readily by bending of the wire and tilting of the aerofoil about the wire. When the fan is switched on, the aerofoil starts to move up and down in front of it. This bodily motion is accompanied by a tilting oscillation. The motion gradually becomes more violent.

That is to say, more than one quantity will be needed to describe the configuration of the system at any instant during its oscillation.

As in the previous example, this self-excitation is caused by a flow of fluid, though now its technical significance scarcely requires explanation. In fig. 59, the semi-circular bar of the previous demonstration model has been replaced by an aerofoil. Up-and-down motion of the aerofoil as a whole (as occurred with the semi-circular bar) is permitted by the long flexible support; it represents motion in one degree of freedom. Rotation of the aerofoil about its own horizontal axis, so that it tilts relative to the oncoming air (thereby distorting a small coil-spring) constitutes motion in a second. There are other degrees of freedom – the aerofoil can move away from the fan by bending the supporting wire, for instance – but they are unimportant. If the fan speed is high enough, the aerofoil moves up and down and at the same time its tilt relative to the oncoming wind, or 'angle of attack', also varies. We have here an example of 'classical flutter', or 'coupled flutter', and it takes place in two degrees of freedom.

The two degrees of freedom that are selected by this model for its motion are available to the wings of an aeroplane. Luckily, however, this does not mean that an aeroplane is *necessarily* subject

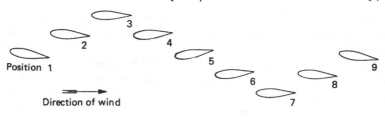

Fig. 60

to this motion. Flutter may also affect the blades of helicopters and the blades that rotate in turbo-machinery. It does, in fact, turn out to be a very extensive subject of study indeed and its avoidance is often a main requirement in design.

As we have seen, this particular form of self-excitation involves two of our system's degrees of freedom. An aeronautical engineer would say that the flutter is 'binary'. Some other forms of flutter that are observable in aircraft structures involve more degrees of freedom and so we have ternary-, quaternary-, quinary flutter, and so on. Before proceeding with a wider discussion of flutter, however, we had better explain the binary flutter that is observed with the system of fig. 59.

The distortions in the two degrees of freedom concerned may be described as those of 'displacement' and 'rotation'. Suppose that both distortions are sinusoidal, that both occur with the same frequency, and that the rotation leads the displacement by a phase difference of a quarter of a cycle. This means that the leading edge is above the trailing edge during the whole of the up-stroke and that the leading edge is below the trailing edge during the whole of the down-stroke. Since the upward force on the aerofoil is increased when the leading edge is above the trailing edge and the downward force increases when the leading edge is below the trailing edge, it follows that the aerofoil is urged on by the airstream during the whole of the cycle.* This is illustrated in fig. 60. A careful examination of the motion of the system of fig. 59 reveals that the required phase difference actually exists while the flutter is taking place, so that the system becomes unstable.

* The reader might like to verify for himself that, if the rotation and displacement were in phase, then the aerofoil would be as much hindered as helped by the airstream during each cycle.

This mechanism of flutter depends upon the fact that a rotation of the aerofoil produces a force of aerodynamic origin which tends to change the displacement, whereas a change of displacement does *not* produce a tendency to vary the rotation by aerodynamic forces. This turns out to be a cardinal feature, not only of this particular problem but of classical (or 'coupled') flutter as a whole. In mathematical language we should say that the *couplings* of the oscillations in the degrees of freedom concerned are *unsymmetrical*.

It would not do to leave the impression that all aircraft are so prone to flutter that they may well fall out of the sky at any moment. To be sure, the early days of flight were marred by failures due to flutter; but ever since then the strictest possible checks have been made to ensure that the trouble does not return.

The onset of flutter involves the selection of a suitable 'flutter mode' with the fulfilment of certain conditions regarding the phase and amplitude relations between the motions in the appropriate number of degrees of freedom. This depends, as it happens, on the air speed* and, for the sake of explanation, let us suppose that this quantity varies. The energy input per cycle is dependent upon the air speed and so is the energy dissipated per cycle by mechanical and aerodynamic damping action. In particular, the ratio of the energy input to the energy dissipated will depend upon the speed. When this ratio is unity it is possible for a steady oscillation to take place, neither growing nor diminishing, and the aeroplane is said to be at a 'critical speed'. The various possible flutter modes of distortion of the aircraft will have their critical speeds and the whole point of flutter analysis is to see that the aircraft is air-worthy in the sense that the lowest critical speed of the system exceeds the highest possible flying speed by a suitable safety margin.

Even if a flutter condition is predicted within the range of operating conditions, all is not lost. Generally speaking, there are three approaches to flutter prevention. The first involves modification of the system itself with the object of uncoupling the motions in the appropriate degrees of freedom (leaving each with a positive damping). Thus the twisting of the aerofoil in fig. 59 about its spanwise axis can be made largely independent of the vertical displacement of that axis. This requires, in fact, that the spanwise

* And also on the air density and the temperature.

axis shall be placed at a particular position in the aerofoil and that the mass distribution of the aerofoil shall satisfy a particular relation.

A second general method that may be used is to increase the natural frequencies by raising stiffness/mass ratios within the structure. This helps because the energy input per cycle during flutter is almost independent of frequency, whereas the energy dissipated per cycle is proportional to it. Thus, for a given aircraft there is some value of frequency such that flutter at a higher frequency is not possible. In so far as the aerodynamic forces draw the flutter frequencies away from the natural frequencies, there is a strong indication that the higher the natural frequencies the better.

These two techniques are the usual ones adopted by aeronautical engineers and are of a highly specialized nature. We shall therefore pursue them no further. The third technique of flutter prevention will not *always* work when aerodynamic forces cause a system to perform a coupled flutter; but we shall mention it in more detail because it is commonly used to prevent self-excitation of other types of system when coupled flutter is not involved. The technique is to insert artificial damping in the system.

By increasing the damping of a system which is subject to coupled flutter – that is, by augmenting the dissipation per cycle of oscillation with a given amplitude – one can *usually* raise the system's critical speed. If the bearings in the system of fig. 59 were sticky (being immersed in oil, for instance) the motion would be suppressed. In fact this demonstration apparatus only works because it has been very carefully constructed to be as free as possible from friction.

This is nearly always the case. The onset of self-excitation in a system can usually be forestalled by the addition of damping in some judiciously chosen way. The aeronautical engineer's difficulty with aeroplanes would become less severe if aeroplanes had not got their characteristic shape; it simply does not permit him to add damping in sufficient quantities where he would like to. Some forms of self-excitation *can* be tackled in this way, however, and we shall mention a few later.

To digress for a moment, let us contrast this last method of suppressing self-excitation with those mentioned earlier for curing resonant forced oscillation. When the vibration was a forced one, the best thing to do was to change the offending natural frequency of the system concerned by de-tuning – either upwards or downwards. Increased damping was then more of a palliative. It is resonable to ask how an engineer can know which of the two cures he should try in any given case. Frequency is the key. If the vibration frequency always coincides with that of some disturbance, then the motion is forced and de-tuning is called for. But if the frequency is not determined in this way by some outside agency – and often it will be near a natural frequency – then self-excitation is indicated and de-tuning may not help as much as damping will.

Many forms of flutter have been observed in aircraft. To describe some of them in words is not easy, as when the aircraft concerned has delta wings. Strictly, a flutter will involve motion of the complete aircraft structure, but it is convenient to ignore this in classifying types of flutter and to concentrate on the aircraft components which play a chief part in the motion. One of the earliest observed instances of flutter – and a particularly easy one to describe – was 'anti-symmetric elevator flutter', by which the elevators moved in opposition like scissors.

It must be obvious that flutter analysis is an essential part of aircraft design and that no aeroplane could possibly be allowed to leave the ground without due attention having been paid to it. It is probably true that more money, more talent and more thought have been lavished by engineers on this particular aspect of mechanical vibration than upon all others put together. Each new aeroplane has its attendant flutter problems and, from what has already been said, it should be clear that increase of air speeds is attended by difficulties. It might be added that supersonic speeds, and speeds which approach the 'thermal barrier', give the aeronautical engineer a good deal to think about in this respect. Rockets and space vehicles also give rise to severe vibration problems and these, by and large, are due to forms of self-excitation.

To digress briefly once again, it can now be seen how vital the phase difference is between displacement and rotation in a simple binary flutter. Fig. 61 may be contrasted with fig. 60; it shows the sequence of positions of a rudimentary stabilizer vane whose purpose is to *reduce* rolling oscillations of ships under way at sea. The 'displacement' is an outcome of the rolling and the rotation is imposed by a machine that is controlled by a sensing-device in the ship.

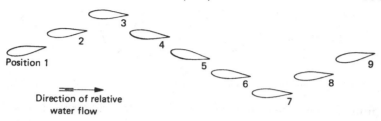

Position 1

Direction of relative
water flow

Fig. 61

The idea – so plausible and so firmly held by so many who should know better – that progress is made by pure scientists having ideas, and engineers making them work, is ludicrously superficial. A successful flutter analyst has to make important decisions as a physicist, as a mathematician and as an engineer. Even a mere description of an aircaft's distortion in a useful numerical form requires a thorough grasp of complex mathematical techniques. Rapid progress in *any* branch of engineering – not just flutter analysis – becomes possible when those who are engaged on research can cross the ill-defined frontier between pure and applied science without realizing that they are doing so.

4.3 Limitation of self-excitation

In the examples of self-excitation which have been mentioned so far, the cause of the self-excitation is hardly affected by the amplitude of the motion. This means that there is a constant tendency for the motion to grow and, unless this is nullified in some way, it will lead to very large swings and possibly to destruction of the system concerned. Many monuments exist to self-excitation in the form of structures which have been utterly broken, yet it is a matter of common experience that total destruction does not always occur.

The dry friction in the hinge of a door may cause the door to squeak, and the motion is a self-excited one since the door is only pushed steadily in one direction. Yet the hinge does not fall apart because the vibration tries to reach infinite proportions. Equally, when a violinist draws his bow across a violin string his instrument squeaks and, although his violin is subject to a self-excitation, it does not fall apart. If oil is placed in the cups at the bottom of the

Fig. 62

springs in the system shown in fig. 56, it is found that the bar will readily bounce up and down in the airstream but its motion will not become as violent as before. In other words this self-excited motion, too, can be limited. (By *complete* immersion of the springs in oil, using longer cups, we could perhaps prevent the motion from starting.)

When the motion is limited in this way it is clear that its cause is something whose effect decreases as the amplitude increases, and so the cause of the vibration is eventually nullified. A steady state is reached in which the energy input per cycle is equal to the energy dissipated per cycle within the system. The vibration therefore acquires a steady waveform like the trace shown in fig. 62, which is a record of lateral displacement obtained during an experiment on the yawing oscillations of a four-wheeled trolley running along a track.

Just because self-excitation often becomes limited we cannot banish it from mind, since the motion can still be both unpleasant and dangerous. On the other hand, the limited motion may be just what we want. This is so with the violin (and for that matter with wind and stringed instruments in general) and with such devices as clocks and electric bells.

Where the cause of self-excitation does not depend for its existence on the amplitude of motion, the oscillation can be described in mathematical terms by means of 'linear equations with constant coefficients'. The exact mathematical significance of this need not concern us here. It is worth remarking, though, that these equations are of a relatively simple kind – although the book-keeping associated with them can become extremely cumbersome if the system concerned has many degrees of freedom. Where the cause of self-excitation depends upon the amplitude of motion it is naturally to be expected that the equations will be more complicated. They are no longer linear with constant coefficients and are described, loosely, as 'non-linear'. Matters are then very different and the equations themselves can be extraordinarily difficult to solve; the equations raise mathematical problems in their own right.

This does not mean that the engineer must always concern himself with the purely mathematical problem of a limited self-excitation. He is usually concerned that there shall be no oscillation at all, whether limited or otherwise. It follows that he may be able to study his problem by means of a 'linear' theory, so that only the process of building up is analysed, while the subsequent limitation by extraneous effects is a matter of academic interest only.

This is the significance of the previous demonstration with the system of fig. 56. If oil is placed in the two cups the motion will be limited. Now for small motions the effect of the oil is likely to be small; one, or perhaps two, turns of the springs will dip in and out of the oil as the motion builds up. But as the amplitude increases this 'one or two' will be increased to more turns and the effect of the oil will begin to show itself more clearly. If our purpose was to prevent all motion of this sort, and to do so by means of damping, we should have to pay attention to the amount of damping within the whole of the system that executes small vibrations and not just to the damping of the oil in the lower cups. That will decide whether or not oscillation can begin at all.

In a perfectly valid sense, the device shown in fig. 56 (with oil in the cups) is not the same *system* when it executes large oscillations as it is when the oscillations are small. This is because one of its characteristics – that of damping – is altered during the large motions. This does not merely mean that damping forces are larger for the larger motions, but that they are *disproportionately* so. The system has a 'varying characteristic' and this possibility is something which will occupy our attention later on. All that need be said at this stage is that the limitation of self-excitation can often be regarded as an extraneous matter which is not of immediate interest. A frontal attack on a self-excitation problem as one of a system with varying characteristics is usually extremely difficult; but, happily, it is not often necessary for the engineer to conduct such an investigation.

4.4 Some practical self-excited systems

A system which is to vibrate by self-excitation requires a source of energy and also some physical process for extracting energy from

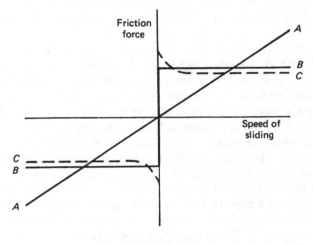

Fig. 63

that source and converting it into a vibration. In the remainder of this chapter we shall disregard the subsequent process of limitation which often comes about. In other words, we shall only consider the linear problems of the build-up of motion, and the fact that the vibrations subsequently become limited will not concern us.

Before going any further, however, we must clear up a matter concerning damping. The damping (or resisting) force which would be exerted upon, say, a knife blade which is made to 'cut' through treacle could be represented by a curve like A in fig. 63. The greater the speed, the larger is the resisting force and, if the direction of motion is reversed, so is the direction of the resistance to the passage of the knife. This illustrates the sort of damping that is supposed to exist in simple vibrating systems. The damping is, so to speak, a property of the system, being specified by the slopes of speed-friction curves (like that of curve A in fig. 63). In the remainder of this chapter, we shall discuss some examples of self-excitation which can be explained, even if only roughly, in terms of the 'viscous' form of damping which is exemplified by curve A.

We may notice, however, that self-excitation may actually depend on the existence of damping, though damping of a rather different sort. Curve B illustrates the behaviour of two dry surfaces rubbing over each other. As before, the friction forces change

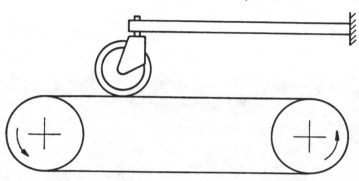

Fig. 64

direction when the direction of rubbing is reversed, though the magnitude of the force is sensibly constant. Actually, dry friction is sometimes more closely represented by curves like the broken line C in fig. 63. Now the essential point, here, is that damping of the types B and C is not that of a system with constant characteristics. It is not possible to specify the friction force in terms of a single quantity that is forever associated with the system and relates the force to the velocity by direct proportionality. This admittedly seems a fine point at the present stage, but, as we shall find in chapter 6, it is an important one. As a matter of fact, we have already referred once or twice to systems – the squeaking hinge and the violin string – whose damping is provided by dry friction.

Turning our attention now to systems whose behaviour in self-excitation can be explained in terms of viscous damping, we may note that, although fluid flow has been used as the source of energy in some of our discussions, it is by no means the only cause of self-excitation. Another type is illustrated by the apparatus shown in fig. 64 which has a castored wheel mounted at the end of a flexible strip of metal. The track on which the wheel runs is a sandpaper band and as the track speed is increased, there comes a stage at which the wheel commences to wobble violently. Similar behaviour is sometimes observed on tea-trolleys and on the trolleys that are used for shifting mail bags in railway stations. As demonstrated by this apparatus, the phenomenon is evidently to be classed as 'binary', to use the language of the aeronautical engineer. There is evidence to show, however, that when the castored wheel has a

Fig. 65. The zig-zag trail left by the nose-wheel of an aircraft which crashed on landing as a result of the shimmying motion. (Courtesy U.S.A.F., Dayton Ohio.)

more readily deformable tyre, the motion is essentially more complicated.

This particular problem is not the genteel, and mildly entertaining thing that is suggested by the tea-trolley, as fig. 65 shows. The undercarriages of some aircraft have given trouble because the nose-wheel is prone to self-excitation. This is still a design problem whose solution is imperfectly understood. The kinetic energy of the moving aircraft provides the necessary energy supply, and the binary system of fig. 64 gives a crude idea of how the motion occurs. A related problem has arisen with trailers which persist in yawing when they are towed behind cars. It has also arisen with Boswell, whose portrait appears in fig. 66.

The whirling shaft apparatus of fig. 40 can be adapted to demonstrate self-excitation. If the previous steel shaft is replaced by one which has a greatly increased internal damping – such as a shaft with a metal sleeve shrunk onto it – the previous behaviour is fundamentally changed. As the shaft is speeded up, the first critical speed is approached as before. But at some higher speed still it ceases to be possible to suppress the vibration; it just goes on and,

Fig. 66. Boswell. He tries to imitate a nylon oil barge (see fig. 2) when pulled along, by bending at his hinges.

no matter what the rotational speed, the shape is that of the first mode.

First we notice that the necessary energy can be drawn from the motor. As for the mechanism by which the energy is siphoned into the vibration, this is to be found in the extra friction that goes round with the shaft. The mating surfaces rub against each other and set up much greater friction forces than do the elements of the plain steel shaft. Notice that the motion is *not* caused by any defect of bend or of unbalance in the shaft. Just why this instability does come about need not concern us here, though we may notice that, while the motion is dependent on the existence of dry friction, this is one of those cases in which viscous damping (if it could be arranged) would produce the same result.

It would be misleading, though, to say that this is an example in which increased viscous damping causes instability, since this particular instability is not caused so much by increase of damping as by increase of the *ratio* of internal to external (i.e. environmen-

tal) damping. If we were to increase the external damping sufficiently, we could stabilize the shaft carrying the shrunk-on sleeve.

Before we leave this question of the self-excited rotating shaft, let
us again refer to the rotor in fig. 41. For it will show the sort of
problem that engineers can run up against. Any good engineer
knows that he must never allow himself to become too alarmed by
theoretical considerations, since technical achievement often
demands a measure of courage and theory can be rather a Jeremiah.
On the other hand it would be foolish to disregard the predictions
of theory – for after all, the theory is drawn up to provide
predictions. Let us, then, think about a point that is enough to
alarm any good engineer.

The steel forging of an alternator rotor carries strips of copper
which are carefully covered by insulating material and are embed-
ded in slots running along the length of the rotor. The wrapped
strips are capable of rubbing against the sides of the slots during
rotation. Now if this should cause one of these monsters to behave
like the shaft carrying the sleeve, the results would undoubtedly be
serious because the shaft would then rotate in such a way as to be
liable to fatigue. With the sleeved shaft the instability becomes
possible soon after the shaft passes through its first critical speed. A
large alternator passes through its first critical speed well below
1000 revolutions per minute. At 3000, it is running in the neigh-
bourhood of its fourth critical speed and has thus run well into the
speed range in which instability is possible. If an alternator rotor
breaks, pieces of metal weighing several tons get thrown hundreds
of metres, sometimes passing through substantial brick walls. Of
course the outlook is not quite as bleak as it would seem at first
sight, though there is plenty of room here for speculation.

Some trains rock violently as they travel along a track and one
way of starting an argument is to try to explain why this is so. The
problem is so perplexing that an international competition was held
in the mid-1950s in an effort to find a solution.* The key to the
matter was eventually found where, apparently, no one in the
railway world had thought of looking – in self-excitation. Now an
oscillating railway train is a perfect nightmare from the vibration
analyst's point of view; it is an obstinate assembly of unknown (and
varying) stiffnesses, masses, clearances and frictions and a great

---

* It failed and, in the light of the rules, it did so predictably.

Fig. 67. A model 'railway vehicle' which oscillates across its track when caused to run above a certain speed.

deal of work has had to be done by railway engineers before high-speed trains could be operated safely on a regular basis.

Instead of plunging into the details of this peculiarly difficult oscillation problem, let us consider the rudimentary 'railway vehicle' that is shown in fig. 67. It possesses two four-wheeled bogies that are spring-mounted with respect to rotations about their vertical axes. The model runs (on rubber rolling surfaces) down an inclined metal track. If the running speed is high enough, the flat-car becomes unstable and the flanges of its wheels commence to clash against the sides of the rails. This is presumably a self-excited motion since no periodic disturbance is applied and there is a source of energy to be found in the work done by the weight of the vehicle as it drops down the incline.

The motion can be roughly described in terms of the forces brought into play by the small distortions of the wheels and rails at their points of contact. Moreover, these forces have roughly the same mathematical form as viscous damping forces (though of an admittedly unusual kind). This model actually behaves as an 'unsymmetrical' system like the aerofoil of fig. 59 and its critical track speed can be raised (so postponing the onset of instability) by augmenting the stiffnesses of the bogie suspensions about the two vertical axes. Damping rotation about these axes has very little effect.

4.5 Self-excitation from fluid flow

Some examples have already been given of oscillation in solid systems which is brought about by the flow of a surrounding fluid. Let us now consider some other examples of self-excitation arising from the flow of a fluid.

Fig. 68 shows a tall steel chimney which has, in fact, no fire-brick lining. Steel chimneys of this sort are much less expensive than brick structures of comparable size and have other attractive features from an owner's point of view. When this particular chimney had been erected it was found to sway in the wind. The motion was not caused by puffs of wind but by a steady breeze. It was a consequence of the periodic shedding of vortices, first from one side of the chimney and then the other. These eddies caused pressure fluctuations on the sides, and hence the swaying. And once the swaying had started it was largely self-maintaining.

In certain other chimneys this same phenomenon of vortex-shedding has caused vibrations of a rather different sort. The motions have been described as 'ovalling', being such that the axis of the chimney remains vertical while the plating executes 'breathing' motions. It is an awesome sight.

This shedding of vortices from alternate sides of an obstacle placed in an airstream has been the subject of much research, though it has never been properly explained. It has been quite clearly established, however, that the wake on the downstream side forms a sort of 'tail' which swishes from side to side.

As usual, this problem of swaying chimney stacks is much more complex than the simple explanation might indicate. First of all the swaying is started by near-coincidence of the vortex-shedding frequency with the first natural frequency of the chimney. That is to say the motion is to some extent one of resonant forced oscillation. But the motion is also self-excited in that, once it has started, the motion (to some extent) dictates the frequency with which the vortices are shed.

One way to stop a chimney like the one in fig. 68 from swaying is simply to guy it (as with the stack shown) and preferably to put dampers in the guys. Alternatively, of course, one can obtain the damping by putting in the fire-brick lining. This particular vibration problem is interesting in that there is a completely different cure available; for the mechanism by which the exciting force is

Fig. 68. A steel chimney stack which swayed in moderate winds as a result of vortex-shedding. The motion was prevented by the guys which can be seen in the picture. (Courtesy U.K.A.E.A.)

Fig. 69. Helical 'spoilers' attached to a chimney to prevent it from swaying in the wind. These spoilers modify the process by which vortices are shed on the down-wind side of the stack. (Crown Copyright.)

produced can be destroyed. Helical spoilers (or 'strakes') may be attached to the side of the chimney as shown in fig. 69 and these have the effect of breaking up the vortex pattern so that no clearly defined excitation is applied to the chimney wall.

Vortex-shedding has accounted for several interesting vibration problems. For instance the periscope of a submerged submarine which is under way will give a blurred image if it is subject to oscillation like the chimney. Again, most seamen know that it can be extremely difficult to tow a body from a ship, particularly if the speed is high or the wave height is significant. The body may veer all over the place and the towing cable often shows a marked inclination either to tie itself into knots or to part company as a result of snatching. One of the main difficulties is 'strumming' of the cable caused by eddy-shedding and this has sometimes to be prevented by threading on plastic stabilizer vanes which act like weathercocks in the flow and stop the unwanted eddy formation (and which are a perfect nuisance when the cable has to be winched in or out). The vanes act as a sort of 'splitter plate' as indicated in fig. 70.

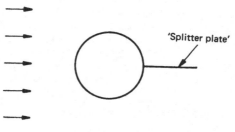

Fig. 70

A case has been reported of large waves in a cylindrical steel storage tank which had been filled with water. The motion had been caused by the wind blowing on the outside surface of the tank and so shedding vortices.

Transmission lines, too, have been found to oscillate in the wind as a result of vortex-shedding. This presents a real problem because it causes high stresses at the supports and may lead to breakage. It is not possible to destroy the mechanism of shedding because any projection from the surface of the line leads to corona discharge, which is pleasing to watch at night but expensive electrically. We mentioned this particular problem earlier, and fig. 44 shows the vibration dampers that are used to prevent fatigue of the lines at their points of support.

Undoubtedly the most famous self-excited oscillation of recent times, however, is that of the Tacoma Narrows Bridge that failed in the autumn of 1940 after only a few months' service. Fig. 1 is a picture taken during that fateful oscillation. Eddies were shed by the structure which carried the roadway; this was not a cylinder but was shaped like a letter I placed on its side. After a great deal of research this bridge has been rebuilt with important modifications. As might be expected, the modifications have not involved the insertion of extra damping so much as alteration of the surfaces presented to the wind. In this way the mechanism by which the excitation arose has been destroyed.

Naturally a great deal of attention has been paid by structural engineers to the Tacoma mishap and the likelihood of its being repeated is remote. This does not mean that wind-excited oscillations in suspension bridges can be forgotten, however, and an interesting instance of it occurred in a different form during the construction of the towers of the new bridge over the Firth of Forth.

Fig. 71. The North Tower of the Forth Road Bridge which sometimes swayed in the wind during erection, before catwalks were suspended. (Courtesy A.D.C. Bridge Co.)

After completion of the foundations, construction of the North Tower (shown, completed, in fig. 71) proceeded without a hitch until a height of about 120 m was reached. The tower, which had withstood winds of up to 36 m/s then began to oscillate in moderate winds of between 9 and 13.5 m/s. This interfered with work at the top. The state of affairs persisted until the full height of 152 m was reached, at which point the motion sometimes built up to an amplitude of 1.25 m at the top and its frequency was about 0.25 Hz. It was sufficient to cause the lowest horizontal joints between the welded box sections to open and close.

The amplitude of this rocking motion was reduced substantially (to about 15 cm) by installing damping guys, so that work could proceed normally. The saddles were then mounted at the top and catwalks were suspended along the line of the main cables to enable a start to be made on their construction. The damping effect of the catwalks was sufficient to nullify the tendency for the tower to rock so that the guys could be dispensed with. The oscillations of the free-standing tower were shown by calculations to be a purely temporary phenomenon occurring during erection under certain very special wind conditions and cannot occur in the completed bridge.

One of the limiting factors in the design of jet engines is the ability of the blading in the compressor to withstand the rough treatment that it receives. The blading is subject to various kinds of vibration of which a form of flutter – 'stall flutter' – is one. This type of flutter is not caused by the *regular* shedding of eddies but is more akin to the phenomenon described in fig. 57. Now to get a fluid to run smoothly from a region of low pressure to one of high (as must be the case in the compressor) is a matter of considerable difficulty. The trouble is that the fluid does not readily adhere to the surfaces over which it is supposed to run but has a tendency to break away and form a disordered wake of eddies.

The aerodynamic conditions within a compressor are difficult to elucidate and the conditions under which the blading is made to vibrate are still a matter of speculation. One thing seems fairly certain, however, namely that increased friction within the material of the blading is likely to be beneficial. It is for this reason that friction dampers are used at the blade-roots; sometimes, even, fibre is used in the manufacture of blading (rather than the high-grade alloys that are normally employed).

Occasionally, flexible shafts which run at high speed in plain journal bearings whip to and fro on account of the behaviour of the oil film upon which they ride. Here again is a self-excitation, the source of energy being the driving motor. The mechanism by which the oscillation is brought about is still obscure. Often there is no real alternative to using plain bearings and their advantages are many; but a high-speed rotating shaft like that shown in fig. 42 would be a menace indeed if it were to execute a violent 'oil whirl' for long. It might be added that oil whirl of this nature can be cured, though this is accomplished on the basis of experience and little more at present.

In every case we have discussed in this section the body whose motion is of interest experiences a varying distribution of fluid force at its surface, that variation depending on the motion of the

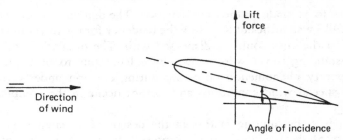

Fig. 72

system. The fluid flow that causes a solid system to commence oscillation need not surround the structure concerned, however, but may flow through it. Thus oil pipelines have been known to vibrate spontaneously and a rubber tube hanging from a running water tap will sometimes flop about in a messy way. This leads us to an interesting and rather unexpected point.

Thinking in abstract terms for a moment, consider a solid body in contact with a flowing fluid. We have seen that the fluid force exerted on the body is determined partly by the body's motion. In the interests of simplicity we have adopted the conventional assumption that the fluid force at some instant is dependent on the body's motion *at that same instant*. Now this is no doubt a very accurate assumption where the flowing fluid is conveyed by a flexible pipe and causes that pipe to vibrate. But in general this assumption is not strictly justified because the flowing fluid cannot adjust immediately to differences of geometry of flow. In other words the fluid force distribution exhibits a sort of 'memory'.

This idea of memory is potentially important, so let us illustrate it simply by using an example in which it is likely to be more significant than it is in pipe flow. Whereas we previously considered an aerofoil that could move in an airstream (see figs. 59 and 60) let us suppose that it is held fixed in an air flow as shown in fig. 72. The flow round the body is such as to apply an upwards force to it, as well of course as a drag force in the direction of the wind. For small angles of incidence this lift force is proportional to the angle of incidence. If the angle were *suddenly* changed, then there would be a proportionate alteration of the lift – *eventually*. The point is that the flow pattern around the aerofoil would have to change and

Fig. 73. A coffee pot which slowly rocks on its rounded bottom when it is placed on a hotplate. (Courtesy U.C.H.M.S.)

this would entail the shedding of a vortex off the trailing edge. The vortex would be carried away downstream and all this takes time.

The argument can be taken a step further. If the aerofoil moves continuously in the wind (though always with a small angle of incidence) its lift at any instant is not what it would grow to be if its configuration were suddenly 'frozen'. Some recent research has had allowance for this effect as its object, though not so much for aerofoils as for ships moving through the water.

Not only in this book, but in engineering generally, self-excitation problems are thought of as relating to a huge collection of isolated phenomena, all requiring individual attention. Fig. 73 shows a coffee-pot whose home is in University College Hospital Medical School. There it habitually rocks back and forth on its slightly rounded base while it is kept warm on a hotplate. The source of energy is the heating element and the influx of energy takes the form of *heat*. An explanation of what makes the pot rock back and forth would demand an understanding of what goes on in the fluid within the pot, and it might be added that in this case a liquid in the pot need not be boiling and can, in fact, be cold. Apart from providing excellent coffee, the pot emphasizes that the physical phenomena which cause self-excitation can be obscure.

5

Shocks and waves

Neath this blow
Worse than stab of dagger —
Though we mo-
Mentarily stagger
In each heart
Proud are we innately —
Let's depart
Dignified and stately.

Any 'everlasting', periodic excitation – even if its waveform is complicated like that of fig. 8 – can be broken down into a series of sinusoidal components. The motion that each of these sinusoidal components produces can then be analysed, being of the type that we met in chapter 3. The question naturally arises – 'What happens if the excitation is not periodic?' An earthquake, for instance, may shake a house for a short time.

In this chapter we shall discuss the oscillations of systems which are shaken by exciting forces that are not regularly repeated. We shall therefore encounter the particular – and technically very important – problem of 'shock'. When, as in shock excitation, the duration of the disturbance is limited and the subsequent motion dies away, it is common to refer to the motion as 'transient vibration'.

Some transient vibrations strongly suggest the idea of 'waves' of deformation transmitted through the material of a system. As we shall see, these waves are by no means incompatible with the vibrations that we have been studying, and they sometimes provide a clearer picture of what is happening.

5.1 Transient vibration

The characteristics of a system in free vibration are most important in deciding what happens when that system is subjected to a transient excitation. If a gong is struck with a padded striker there is a short period during which the striker is actually in contact with the gong. After this the gong is left vibrating freely on its own, emitting the sound which is familiar to us all. The free vibration is executed by the gong in its various natural modes, each with its own particular frequency. The motion in these various modes gradually dies out because of the friction forces within the gong – it is suspended in such a way that little damping arises at the supports – and on account of energy radiation in the form of sound. A somewhat *different* response can be obtained from the gong by striking it with, say, a wooden rod rather than its padded striker. The free vibration is then different because the relative intensities of the vibrations in the various modes are different.

Forms of transient excitation which occur in engineering systems are very numerous and varied. The suspension of a car, for instance, is subjected to it when the brakes are applied, or the vehicle is accelerated. So, too, is a door which strikes a door-stop.

If a building is close to the centre of an explosion, then the blast wave may demolish it and its behaviour would be outside the scope of this book. On the other hand, if the structure is far enough away it will merely be shaken. When the atomic bombs were exploded over Japan towards the end of World War II, these monstrous weapons demolished large numbers of structures, with the startling exception of tall chimneys, many of which remained standing. Fig. 74 is a picture of Nagasaki, looking towards the point – about 1.5 km away behind the centre group of factory chimneys – above which a weapon exploded. It shows several chimneys still standing despite the general devastation around them. The reason why this was so will become clearer later on.

Transient excitation by a blast wave bears little resemblance to that set up by a vehicle crossing a bridge. Suppose that the vehicle is a motor-car, whose mass is negligible compared with that of the bridge. The structure experiences a constant force – the car's weight – which moves across it. It is clear that the bridge will deflect under the car's weight. It is equally obvious that the deformation will vary since the position of the car is not fixed. The

Fig. 74. The devastation of Nagasaki looking towards the spot over which an atomic bomb exploded. Notice that chimneys were left standing while other buildings collapsed. (Crown Copyright.)

transient vibration of the bridge will be determined in part by the speed with which the car crosses it. On the other hand, when a locomotive crosses a bridge it contributes much of the total mass that is set into oscillation. This complicates the analysis quite considerably (and things are not improved by the regularly repeated thumping that the bridge receives from a steam locomotive as a consequence of the form of its driving mechanism).

Aeroplanes are subjected to transient loading. The most obvious example is the bump that is administered when they first touch down on landing. An aircraft that flies through a gust of wind or a region of turbulence also experiences this sort of excitation.

As we have seen, a rotor like that shown in fig. 41(a) is driven at 3000 revolutions per min by a steam turbine, in order that it shall generate electricity with the standard frequency of 50 Hz. The portion of the rotating system shown in that figure is, in effect, a very large electromagnet having a north and a south pole. The current which energizes this magnet is fed into the rotor through slip-rings. The rotor runs in a stator that is composed of an iron frame carrying coils. Electricity is generated in these coils, being then passed to the transmission lines. Whenever there is a sudden change of the electrical load demanded from the stator – and a short-circuit is an extreme case – the rotating magnet is subjected to a transient torque. This torque, whose variation with time depends upon the nature of the change of load, causes the rotating system to take a sudden extra twist, so that the turbine commences to oscillate in torsion relative to the rotor. The whole motion takes place about the steady running configuration at the operating speed.

In the last few years engineers have been studying problems of aperiodic loading, as much with the intention of finding the true nature of the loadings as with studying their effects. This branch of research is a steadily growing one and a moment's thought about all the aperiodic fluctuating forces that can be applied to ships, aeroplanes, cars, bridges and structures in general will show that this is an immense undertaking.

5.2 Slow and sudden transient excitation

We have begun to see that the problem of transient excitation is fundamentally a rather complicated one. One of our troubles is that it is difficult to find a method of generalizing results. Some generalization is possible, however, on the basis of the abruptness of the transient loading. In this discussion we shall leave aside all questions of the modification of systems to alleviate the effects of shock loading and just address ourselves to the problem of the rate at which the shock loading is applied to the system.

Consider, first, 'slow' transient loading by a force which does not make abrupt changes. All of its major changes of magnitude occupy periods of time that are much greater than the time of one oscillation of free vibration in the mode concerned. In this case the deflection of the system occurs as if the force were applied statically, and the deflection at any instant is that which could be expected from the prevailing static force. That this result is of considerable technical importance can be illustrated by means of an example from civil engineering.

Suppose that a car travels at a constant speed along a straight road and that it then enters a circular arc of turn which it

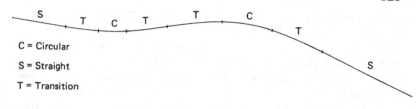

C = Circular

S = Straight

T = Transition

Fig. 75

subsequently leaves, to run off at a tangent again. At the instant when the car enters the curve, a sideways inertia force is applied to it, tending to roll it outwards. That is to say the car is suddenly thrown sideways and will thereafter commence to oscillate on its springs until the dampers kill the motion. On leaving the curve the car again swings upright; but now it swings about its original position of equilibrium, having been subject to an abrupt removal of the force.

This would all be rather uncomfortable (although, of course, drivers would instinctively minimize the effect by moving across the traffic lane). In fact curves in roads are not laid out in simple circular arcs for this reason, and care is taken to blend the corner with the straight by means of a 'transition curve'. The effect of the transition curve is to make the sideways force grow more slowly from zero to its maximum value. If the period over which this takes place could be made much greater than the lowest period of rolling motion of the car, then the driver would not feel an oscillation but simply a slow change of the car's attitude on its suspension.

While each bend in a motorway like the M1 is composed of a circular arc and two transition curves, it is usually so slight that this is not made apparent by aerial photographs. A fairly severe bend has been made in the Doncaster Bypass Motorway, however, the line being as shown in fig. 75. But even in this case the difference between the circular arcs and the transition curves is very slight.

If a car runs over an obstacle, its wheels are made to perform a vertical motion of limited duration. This is not communicated directly to the passengers because of the springing, both in the tyres and between the axles and the chassis. The wheels may be said to suffer transient displacement excitation. The springs of the car, having been distorted in this way, now begin to cause the passengers to bounce up and down in a free vibration.* Again the question of 'slow' and of 'sudden' loading arises. If a car crawls over a hump bridge then the conversation of the passengers need not be disturbed. But if the car should speed up and go over the bridge too rapidly, the consequences may be determined partly by when the occupants last had a meal; the excitation is no longer slow.

At the other extreme from 'slow', we have 'sudden' loading. Here the applied force fluctuates so quickly that the whole process of loading is over long before a complete cycle of motion in the mode concerned can be executed freely. This is sometimes known as 'impulsive' loading. Mechanics tells us that the force imparts momentum to the system, the momentum being equal to the impulse of the force. The system now sets off on a free vibration with the velocity that has been given to it before it has had time to budge from its rest position. The free vibration gradually dies away and that completes the process. The system scarcely distorts during the process of loading and only reaches its maximum deflection later on during the free vibration. This is probably what saved the Japanese chimneys shown in fig. 74.

As we shall see, this matter of sudden loading suggests a different method of looking at problems. It leads naturally to the ideas of wave analysis.

5.3 Free stress waves

Once or twice we have come close to the boundaries between research in vibration and research in other subjects closely connected with it. Each time we have glanced over the boundary and then hastily withdrawn. Now we come to a boundary that must be crossed.

Let us consider the transient motion of a system in which we must take account of more than one mode. A good example is provided by a long clothes line, one end of which is attached to a rigid object and the other end held in the hand. If the line is stretched out and the end is very slowly raised and lowered, the line slowly rises and falls. This is the motion that we called 'shaking' in chapter 1. When the hand is stopped, the line stops moving and

_____

* This motion is damped out fairly rapidly, however, by the dampers. In fact if the dampers are omitted from a car the passengers have a very bumpy ride indeed.

there is no residual vibration although, with a very long line, it may be quite difficult to move the line slowly enough for this to be true.

Suppose, now, that the motion of the hand is not as slow as this. Then, while the hand is actually moving, the rope is subjected to a form of excitation and, when the hand stops, the rope continues to vibrate freely. Now this free vibration can take place in one or more of the modes of free vibration of the rope. Thus we could leave the rope oscillating almost entirely in its first mode by shaking the end of the rope up and down at approximately the first natural frequency and then stopping suddenly. This is a technique that we employed in section 2.1. Equally, we could in theory leave the rope vibrating largely in the second, third, . . . mode, by exciting the rope suitably.

Consider, now, what happens if the free end of the rope is given a quick flick whose duration is much less than the first period of the rope. We should expect to leave the rope with a free motion in several modes – possibly quite high ones. In fact what happens is that the distortion that is imparted to the rope is seen to run along the rope in the form of a 'wave'. The wave reaches the fixed end of the rope and is there reflected so that it travels back to the hand. It then becomes reflected again and runs to-and-fro along the rope in this way until it is finally obliterated by the friction of the rope.

This is worth a little consideration. We see that a quick flick of the end of the rope leaves a wave travelling to-and-fro along the rope. But we also know that that free motion is taking place in the various modes of the rope. The mixture of distortions in these modes is not a particularly simple one; all we can be sure of is that it has a large 'high-mode' content. Whereas we can see the wave, we have some difficulty visualizing the motion in the modes.

An alternative piece of apparatus for showing this behaviour is worth describing. We need a system with many low natural frequencies and with a sufficient extensiveness, in the geographical sense, to display motions clearly. A length of steel band may be suspended from a high ceiling with transverse bars attached to it at regular intervals, as shown in fig. 76. The bars may be about 0.25 m apart and, say, 0.5 m long with knobs at the ends so as to increase their massiveness about the axis of the band. Such a metal band may be used as a torsion pendulum. A twist imparted to the band at the bottom will be seen to travel upwards, be reflected and travel downwards again.

Shocks and waves

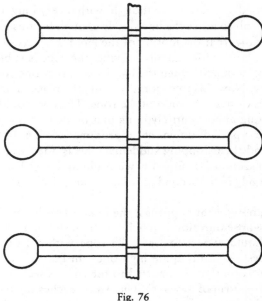

Fig. 76

This idea of a travelling wave is so much simpler than the rather complicated notion of vibration in modes, that we should be foolish to disregard it. This is particularly true under two specific conditions. The first of these is that the terminating boundary of the system that supports the wave is very far from the source of the disturbance. The second is that the system that supports the wave should be sufficiently 'sticky' to damp the wave out before it has got to the far boundary.

In these circumstances, the wave's return will not complicate matters. It is as though the long clothes line were kilometres long so that we could flick the end of it and just watch the wave run away, never to return. This is of course perfectly familiar and is well illustrated by dropping a pebble into a pond. The little wavelets that spread out from the disturbance can usually be relied upon to become dissipated when they reach the bank and are therefore unlikely to return to the point at which the pebble was dropped in. We may speak of 'free waves' being propagated in this era of free vibration after an initial disturbance.

When we start thinking about waves of this sort, we are faced with a number of fresh ideas. One acquires a sort of 'wave consciousness'. For instance one is led almost without thinking into the idea of conveying information of some sort by means of a wave. The clothes line can be made to carry an instruction in the form of a distortion, and so can the torsion pendulum. A rather remarkable illustration of this idea can be had from a very long spiral spring which is hung from a high ceiling down to a point near the floor and which has some suitable object hanging on its end. If this object is taken in the hand and suddenly moved upwards a distance of, say, 0.5 m, the 'instruction' that the object shall stay where it is put is carried up the spring. The object will remain there stationary until a wave has travelled up the spring, been reflected at the top and come down again. Had the far boundary been a very long way away, the object would stay where it was put.

This idea of transmitting information suggests an explanation of the protection afforded by packing an object in foam plastic. When the package is dropped, the impact 'message' from the outside of the packing only reaches the protected object in a highly modified form. The wave is heavily damped.

The 'information' that is transmitted in a wave can take the form of a stress. If the end of a rod is struck with a hammer, a distortion is transmitted down the rod; and the distortion carries a system of stresses with it. Suppose that the rod is in fact a conical one and is made of a brittle material. If the rod is struck at the bottom of the cone the wave sets off (as one of compression) towards the apex, is there reflected as a wave of tension and this tension can cause failure as it travels back along the rod.

Behaviour of this sort can be illustrated rather prettily with a row of pennies that are placed in a well-powdered slot, rather as if they are on a shove ha'penny board. If twenty pennies are placed in the slide, touching each other, and one is drawn back and made to strike the remainder, it will be found that one penny will fly off the far end of the line. If two pennies are drawn back, two will fly off the far end of this line. And so one can go on. It is possible here to think of the pennies as giving rise to stress waves which travel along the lines of the pennies – both the struck and the striking.

The idea of a barrier to waves suggests itself. Suppose, for instance, that a large drop-hammer is used in a workshop. Each time the machine operates, it sets up a free wave in the ground upon which it rests. This wave will shake any other machine, or possibly cause annoyance in other ways. Is there any way, therefore, of containing the free wave and preventing it from spreading out? In actual fact it can easily be shown that a great deal of benefit can be had by mounting the machine on suitable vibration isolators.

Not only will the drop-hammer shake the ground, but it will also impart free sound waves to the air. Can they be suppressed? This opens up the whole question of sound-proofing. There is no doubt that the whole subject of applied acoustics is going to become more important in the years to come.

As we have seen, the idea of a wave introduces certain ways of thinking. Waves are reflected, for instance, and the whole idea of reflection is absent from the discussion of vibration. In certain circumstances, the shape of a wave will change as it progresses through a body. Again, certain waves are essentially associated with travelling along the *surface* of a body rather than through it. The behaviour of free waves as they are propagated through solids and fluids is a very large subject indeed.*

5.4 Forced waves

We have seen that free waves are of considerable interest when they are propagated through an extensive body – that is, a body which is large in relation to the size of the distortion being propagated. This is particularly true if the far boundary of the solid is very far from the source of the disturbance and if the friction within the body is such as to dissipate the wave substantially before it can be reflected and come back to the source of the disturbance. As we shall now see, the idea of wave propagation is much more valuable even than this.

Let us go back once again to the long clothes line that is fixed at one end. Suppose that the free end of the line is given a small regular *everlasting* vibration. If the line were long enough and its motion were sufficiently heavily damped, it would be found that only the end of the rope in the vicinity of the shaking would be in motion. It is as if a wave passes into the rope from the end that is

* The term 'free wave' is sometimes used in a slightly different sense, notably in acoustics. Whereas we have employed it when the excitation is of brief duration and is applied in a limited region of an extensive system, the acoustician finds it useful when only the second of these restrictions is met. He does not mind, even, if the excitation is sinusoidal.

being shaken and travels towards the fixed end; it does not reach the fixed end because it is annulled by the friction before it gets there. This is what happens if a sound is emitted in free air. A vibration is imparted to the air and is transmitted away from the source in the form of a wave. But if one gets far enough away from the source one ceases to hear the sound because it is sufficiently attenuated, having both spread out in three dimensions and also been attenuated by small friction forces.

Under these conditions we have what we might call a 'forced wave' – 'forced' because the disturbance is not of limited duration, but is now periodic. Quite apart from the transmission of sound waves in air, forced waves are very common indeed. A vibrating machine in a workshop, for instance, will set up a vibration in the structure upon which it is supported. This vibration is really evidence of a forced wave which proceeds from the machine out into the structure. It gives rise, for instance, to structure-borne noise. One of the ways of combating structure-borne noise is to interpose anti-vibration mounts between the machine and the structure. Fig. 77 shows a machine that is supported in this way.

Those who have swum under water in a lake may have discovered that the noise of a motor-boat engine can be carried with great clarity over very long distances indeed. The best way of detecting submarines is by the use of sound waves. The submariner who does not wish to advertise his presence must not, therefore, impart sound waves to the water surrounding him by the use of vibrating machinery. Tremendous efforts are made in the construction of machinery for submarines to see that sound waves – forced waves – are not transmitted.

We have seen that the idea of a free wave is in no way incompatible with the idea of free vibration in several high modes at once. With forced waves we are really discussing forced vibration. But this forced vibration is of a particular sort: it is forced vibration whose frequency is much higher than the lowest natural frequency of the system that is experiencing the motion. In other words the forcing frequency is high up among the natural frequencies of the system. When this is the case, it is not possible to pick out any one mode and bring it to resonance. A whole lot of modes are brought to a state of near-resonance at the same time, and this has the effect of localizing the distortion near the point of application of the disturbance.

Fig. 77. A heavy machine tool supported on 'anti-vibration' mounts. By interposing mounts between the machine and the floor, it is possible greatly to reduce the transmission of stress waves away from the machine. (Courtesy Cementation (Muffelite) Ltd.)

It will be remembered that by a 'forced vibration' we mean one that is caused by some disturbance whose existence is not dependent upon whether or not it is able to cause a motion. The disturbance would be there whether or not the system vibrates. We have also seen that the disturbance need not be sinusoidal but can be of a random character. Just the same is true where forced waves are concerned. Intense noise, for instance, can cause a vibration that is transmitted into a structure away from the point at which the pressure fluctuations associated with the noise actually impinge. This is illustrated in fig. 78 which shows an internal rib from an aircraft. The noise from the aeroplane's jets played on the tail unit and set up a vibration. This random motion set up forced waves which were transmitted into the structure. This structural member has suffered serious fatigue damage due to the stresses set up in this way.

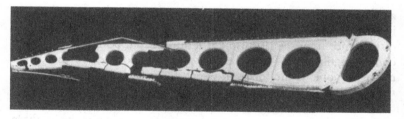

Fig. 78. The remains of an internal member of an aeroplane. The fatigue damage was caused by noise which set up stress waves in the aircraft structure.

The study of random forced waves is only in its infancy. When an excitation like that which caused the failure shown in fig. 78 sets up random forced waves in an aircraft structure, the waves proceed into the structure away from the seat of the disturbance. The waves usually cause damage – if any at all – at junctions between members or at other discontinuities within the structure. This strongly suggests that the damage is not unconnected with the reflection of forced stress waves. This raises an obvious question – 'do forced waves obey the same rules as free ones?'

The answer to this question is that forced waves do obey the same rules as free waves in general, but there is one major difference between the two sorts of waves. If the forced wave is set up by some sinusoidal disturbance, then that wave has a characteristic frequency all of its own. This means that while forced waves can be reflected and refracted in the same way as free ones, they also display certain properties peculiar to themselves. One of these is the Döppler effect, which is that a sound emitted by a moving source has a different pitch (frequency) according to whether the observer is being approached or being left behind by the source. A railway train's whistle has a varying note as it rushes past someone listening to it.

In the next section we shall discuss one particular example of the way in which the possession of a frequency by a forced wave can have profound consequences. For as we shall see, a high frequency implies a large energy transmission by the wave.

5.5 Ultrasonic vibration and waves

We have seen that the notion of a transient leads naturally to that of a 'free wave'. Further, we have seen that the disturbance which initiates a stress wave can be a sinusoidal one, so that the disturbance imparts a sinusoidal motion to all points of the body conveying the wave – a 'forced wave'. Let us now follow this up by discussing one of the forms in which these forced stress waves are used, mainly in liquids and solids.

Quite recently, engineers have acquired a new tool in 'ultrasonics'. Ultrasonic waves are fundamentally the same as audible sound waves, but are of higher frequency. A vibrating surface imparts energy to the air, and the air transmits this energy in the form of a forced wave. If the wave reaches our ears we experience the sensation of sound, provided the vibrating surface moves with a frequency that is between about 18 and 18 000 Hz. Although the waves set up by vibrators at still higher frequencies are essentially the same as these 'audible' ones, we cannot hear them and they are therefore described by the adjective *ultrasonic*. Ultrasonic waves have some remarkable properties, one of them being that they can transmit much more power from one point to another than ordinary sound waves can.

Bats are known to navigate by the use of reflected ultrasonic waves which are produced by special organs that can vibrate with frequencies up to 70 000 Hz. Their ears respond to these high frequencies so that a bat can send out an ultrasonic wave and hear reflections from obstacles in its path. Dogs can also hear some ultrasonic waves: a man blowing a Galton whistle can produce a wave having a frequency of more than 18 000 Hz. While a bystander can only hear a gentle puffing noise, a dog can hear the whistle from distances little short of a kilometre.

Ultrasonic waves can be made more 'penetrating' than ordinary sound waves in the sense that they more readily form beams, like the light from an electric torch. This ability to concentrate the waves into beams becomes progressively greater as the frequency is increased. Whereas a radio loudspeaker vibrating at a few hundred hertz effectively radiates in all directions equally, a vibrating crystal, whose frequency may be as high as a million hertz, sets up ultrasonic waves that can be beamed in straight lines like rays of light. And, like a ray of light, the ultrasonic beam may be reflected by a plane mirror – or focused by a concave one – without much loss of power. In fact sound waves and ultrasonic waves obey all the familiar laws of wave motion and are subject to reflection, refraction, dispersion, interference, diffraction – just like light waves, for instance. But it is far easier to demonstrate and use these properties with ultrasonic waves than it is with ordinary sound because the experiments require only mirrors, lenses, gratings and so forth, of moderate size, the wavelength being so much smaller.

As we shall see, these ultrasonic waves have some very unexpected and useful features. Let us first inquire, though, how the waves may be produced, remembering that a vibrating surface is needed whose frequency (for the wave to be 'ultrasonic') must be at least 18 000 Hz.

Generation is possible by purely mechanical means – as by Galton's whistle. Other mechanical devices include sirens and tuning forks. But mechanical exciters are never used now, since they are neither as efficient nor as convenient as certain electro-mechanical devices.

Perhaps the most common type of exciter is based on the principle of 'magneto-striction': the physical dimensions of a suitable metal body are changed by applying a magnetic field to it. The simplest practical ultrasonic generator of this type is merely a rod of nickel driven at resonance in its first mode by passing an alternating current at the correct frequency through a coil wound round it. This exciter would have a node at its centre and maximum amplitude of longitudinal displacement at each end, one of which would be the radiating face. The amplitude at a radiating face may be about 0.013 mm and magneto-strictors are most useful in the frequency range 5000–50 000 Hz.

Much higher frequencies may be obtained by the use of a suitably cut portion of a crystal, operating on the piezo-electric principle. The crystal, which is usually a small quartz plate or a sliver of barium titanate, changes its dimensions when a potential difference is applied across it. The crystal is suitably mounted and is then driven at resonance by applying an alternating voltage across its surfaces, which are silvered so as to permit electrical contact to be made. A 'thickness mode' is usually employed, in which the two faces of the plate alternately approach and recede from each other. These crystal exciters can be used conveniently at frequencies between 250 000 and 2 000 000 Hz.

Crystal exciters are sometimes used, not in resonant vibration, but in transient. The crystal is given electrical 'jerks' by the repeated application of a high potential across it. One of these jerks sets the crystal 'ringing' like a bell, at its own very high natural frequency. The vibration then dies away, owing to radiation of energy away from the crystal face in the form of a transient ultrasonic stress wave and, to a lesser extent, to the presence of damping. When the vibration has disappeared, another electric pulse sets the crystal off again. This all happens quite quickly – in fact, the mains frequency of 50 jerks per second is usual in Great Britain.

We now know how to get a beam of stress fluctuations which may be either an 'everlasting' sinusoidal effect of ultrasonic frequency or a series of pulses, each having ultrasonic frequency. Such beams have many uses, perhaps the most important of them being in navigation. The waves are generated by apparatus carried in ships and are radiated into the sea. They are reflected from underwater obstructions, from the sea bed or from other vessels, and their measured times of travel give the distances they have covered. We thus use ultrasonics in much the same way as bats do. A form of echo-sounding gear has even been used for the location of shoals of fish.

Let us return yet again to the alternator rotor of fig. 42. It is cut from a single steel forging. Such are the stresses set up in the rotor when it is up to its operating speed, that it would be dangerous if there were cracks or flaws in it. Ultrasonic waves are used to investigate the homogeneity of these forgings before they are accepted; the waves are reflected by cracks and other discontinuities and give much information about the internal condition of the metal.

It happens that the power transmitted across unit area in a compression wave is proportional to the square of the frequency, for a given amplitude. Ultrasonic waves can thus transmit far more power per square centimetre than acoustic waves can. The power transmitted by an ultrasonic beam may be further concentrated by focusing it. This may be done, for instance, by directing the wave along an inverted horn or a tapered piece of metal. By this means it is possible to obtain very intense, high-frequency vibrations such as may be used for fatigue tests *of short duration.*

These vibrations have some remarkable powers. If, for instance, an ultrasonic transducer is used with a tapered metal focusing rod, a 'drilling machine' can be made. A reasonably strong piece of metal attached to the thin end of the taper will be made to vibrate in the longitudinal direction and, if suitably fed with an abrasive paste, it will eat its way through very hard materials like glass and tool steel. In this way it is just as easy to drill a square hole as a round one, since the 'drill' does not rotate.

An attempt to focus an ultrasonic beam in a liquid only meets with limited success. too high a concentration of energy causes an accelerated dissipation of energy by the process of 'cavitation'. The liquid contains large numbers of tiny gas bubbles trapped in it and, when cavitation occurs, these are rapidly expanded and collapsed. The actual process of cavitation is complicated, but its effect is to set up 'points' of very high pressure within the liquid. By no means confined to ultrasonics, cavitation commonly occurs in hydraulic machinery, notably near the surfaces of ships' propellers. It may be thought of as the attempt made by a liquid to prevent a tension from being set up within it.

The highly localized points of great pressure possess some very inconvenient, as well as some useful properties. If they lie near the surface of a metal, they tend to eat away the surface. Indeed the driving face of a powerful ultrasonic generator that is immersed in water becomes pitted because of this effect. On the other hand, this disruptive action may be used to make emulsions of liquids which (like oil and water) will not normally mix. Again, cavitation kills bacteria, and so ultrasonics could be used for sterilization. Yet again, solder may be made to flow on to aluminium – which it will not normally do because of oxidation – so that ultrasonic vibration-induced cavitation makes aluminium soldering possible.

Some of the effects that have been mentioned are illustrated in Fig. 79. A magneto-strictor *A* is shown immersed in water. It produces an ultrasonic beam of the steady vibration type and this is being reflected by the plane surface (a 'mirror') *B*. From the mirror the beam is passed to a concave surface (a 'reflector') *C* which focuses it on a point in the liquid just below the surface. It can easily be shown how narrow the beam is by slightly rotating the mirror *B*, for the beam is found not to strike *C* unless the angle of *B* is carefully set. As we have seen, focusing an ultrasonic beam in a liquid may produce cavitation. This is the case here, and the result is that a small fountain of water is produced, rising from the surface of the water as shown.

There are many other uses to which ultrasonic waves may be put. Ultrasonic therapy has helped sufferers from rheumatism. Again, the scale on teeth may be removed by a dentist using an ultrasonic

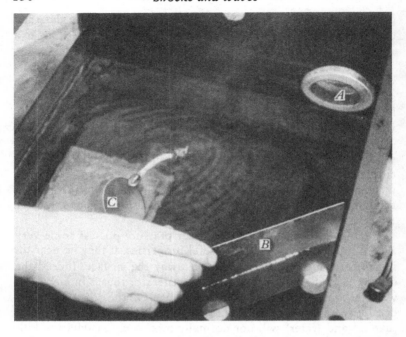

Fig. 79. A small water fountain set up by focusing an ultrasonic beam in water. High-frequency vibration of *A* sets up a wave that is reflected by the plane surface *B* and focused by the concave reflector *C*. (Courtesy Mullard Ltd.)

vibrator; yet again, metal parts may be cleaned very efficiently if irradiated while immersed in a solvent. Ultrasonics have even been used to hasten the maturing process in wines and spirits.

5.6 Transient excitation that is neither slow nor sudden

Let us briefly recapitulate. We have seen what transient vibration is. If the excitation is 'slow' the response is essentially static. If the excitation is 'sudden', on the other hand, the concept of a free wave becomes useful. (We have taken the opportunity of examining *forced* wave phenomena, though strictly that is something of a digression even though the subject is technically an important one.) Now let us consider the more general case in which the excitation is somewhere between these extremes. For some problems of this sort are of immense significance.

Fig. 80. Unless its speed is suitably reduced, a ship may 'slam' in heavy seas. Like this coastal minesweeper on trials, it will lift its forefoot clear of the water and then slap the surface on re-entry. Such may be the violence of the impact, the hull will be left shuddering for a significant period. (Courtesy Royal Navy.)

In seismologically active countries (such as those surrounding the basin of the Pacific Ocean) the study of earthquakes, and their effects on buildings, is a serious matter. Aseismic design has recently acquired very great importance, because Japan – an industrial and heavily populated country with limited resources of fossil fuels – has decided to embark on an extensive programme of building nuclear power stations. Japan is subject to severe earthquakes so that the structures and equipment of a nuclear station have to be very carefully designed so as to avoid a radioactivity hazard to the public, should an earthquake occur at the site.

The sea can subject a ship to transient loading. Fig. 80 shows a coastal minesweeper that is being driven hard in a heavy sea. It will be seen that the forefoot has lifted clear of the water. When the hull strikes the surface on re-entering the water it will receive a blow that is distributed over a length of the bottom and this transient loading will persist as the flared bow 'digs in'. The intensity and distribution of this form of loading are very hard to estimate but its duration will be of the same order as the periodic times of the lowest symmetric distortion modes. Attempts have been made to estimate the responses to this transient loading since the stresses set up may be much greater – perhaps 4 or 5 times greater – than those caused in the random sea when bow emergence does not occur. Here again, the transient loading of this non-conservative system is neither 'slow' nor 'sudden' so that no simplification is available on these accounts.

How can we study transient excitation that is neither 'slow' nor 'sudden'? Now that it is no longer open to us to consider a sinusoidal 'standard excitation', generalization of our study from periodic to transient motion is obviously going to raise some difficulties. Perhaps the biggest difficulty is that, to study transient vibration, we really need the results of experiments that cannot possibly be performed accurately. Before mentioning what these hypothetical experiments are, though, it will be as well to perform quite a simple one.

We have seen that in transient vibration there is first a period during which the excitation is acting and then a subsequent era of free vibration. For the present, we shall study systems that move almost entirely in a single mode. It follows then that the free vibration is merely a subsiding oscillation with a single frequency (very nearly a natural frequency). Now bearing in mind that anything like a systematic study is likely to be rather complicated, let us investigate in a rough-and-ready way what happens if we change the system's behaviour in free vibration. After all, we have plenty of evidence now that the frequencies, mode shapes and damping of a system are profoundly important in other types of vibration. Obviously the characteristics in free vibration are important during the second era of transient loading, but it may be that they are also important during the first (while the transient excitation is actually acting).

Fig. 81 is a sketch of a piece of demonstration apparatus. The apparatus is composed of a desk blower A, to the nozzle of which is fitted an old plate camera B. In front of the camera there is a pendulum consisting of a metal rod to the bottom end of which a plate is fixed, C. When the shutter is operated a puff of air is administered to the plate, causing the pendulum to swing away from the camera. The frequency of free vibration of C can be reduced by attaching a mass to the top of the pendulum above the pivot. With a fixed blower speed and a pre-selected shutter setting on the camera, it is found that the extent to which C swings back is

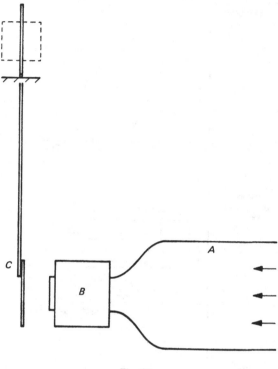

Fig. 81

very much dependent on whether or not the mass is attached to the top of the pendulum.

There are three things that we should notice about this little demonstration:

(1) The massiveness of the pendulum is increased by the attachment of the mass above the pivot;

(2) it is probably true that the aerodynamic force exerted on the plate is more or less the same whether or not the mass is attached to the top of the pendulum;

(3) the value of the ratio (duration of aerodynamic pulse)/(period of one swing of the pendulum) is changed by the attachment of the mass because, although the length of the pulse is more or less the same (being fixed mainly by the shutter setting) the periodic time is greatly changed. Without much surprise we learn

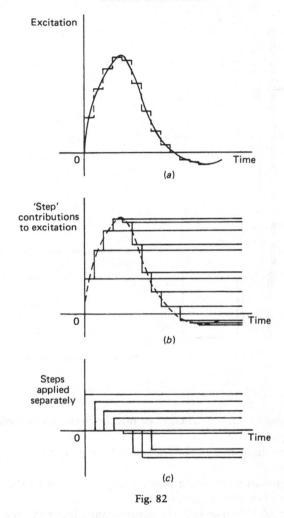

Fig. 82

that, yet again, the 'dynamical personality' of a system determines how that system reacts to excitation.

Fortified by this reassuring discovery, let us briefly think about the experiments that we cannot perform in fact (but whose results we can certainly attempt to calculate theoretically). We shall mention three such imaginary experiments.

Fig. 82(a) shows the time variation of an excitation, like the pulse

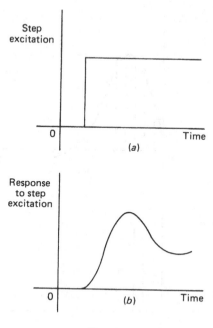

Fig. 83

of force administered by the blower of fig. 81. The curve has been approximated by the dotted steps and we may bear in mind that the approximation could be as accurate as we please if we are prepared to make sufficiently small steps. In fig. 82(b) the emphasis is changed and we see the original curve dotted and the steps drawn in full lines. Taken separately these same steps are as shown in fig. 82(c). If we know the response at any point of interest to just one step of excitation, therefore, we should be able to find the total response to our series of steps by a suitable process of addition. Thus if we could possibly apply a sudden 'step puff' from the blower, as in fig. 83(a), we could measure the variation of the angle of rotation of the pendulum away from the vertical, which variation might be as in fig. 83(b). This idea is fine for theoretical work but, as we can readily see, quite hopeless for experimental.

The same state of affairs arises with the alternative approach illustrated in fig. 84(a). The pulse is divided into vertical 'strips' (and again the division can be as accurate as we please). Let us

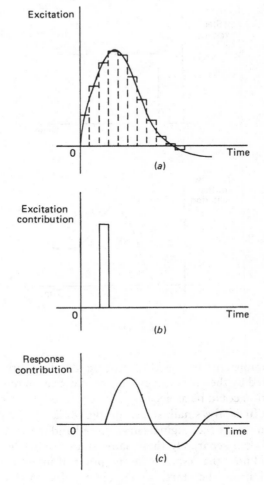

Fig. 84

consider just one of the strips and suppose that we could, indeed, accurately apply such a variation of excitation as in fig. 84(*b*); the response might then be like that of fig.84 (*c*). If we had a suitable process for adding the appropriate responses we could build up the total response to the original pulse.

The third impossible experiment takes us in fact to familiar ground. Take another look at fig. 10 which, it will be remembered shows one cycle of my pulse. If we imagine this one cycle stretched out in time indefinitely, it would become a transient vibration instead of a periodic one. The harmonic content of fig. 10 would still exist, but with the important difference that the change of frequency between two successive components would become very small. In fact the transient pulse would have contributions from *all* frequencies, and not just a set of discrete frequencies. To translate this into the terms of the blower system of fig. 81, if we knew the response (of, for instance, pendulum angle) to a sinusoidal excitation for all frequencies from nought to infinity, then we could discover the response to a transient excitation. We should analyse the excitation into its sinusoidal components and thence synthesize the response. Yet again this is fine for theoretical work but impossible in practice.

Summing up then, we see that there *are* 'standard excitations', even for transient vibration. (And as one would expect, they are related to each other mathematically.) All of them require a judicious addition of responses and this is readily provided by the mathematical technique called the 'calculus'. Even so, it is undoubtedly true that transient vibration is more difficult to estimate and to control than periodic. Being frequently a serious matter technically, this means that devices are often employed which can only be described as 'crude but effective'.

The transportation of delicate apparatus has provided engineers with many problems in the past. How, for instance, can one send a large thermionic valve on a railway journey without presenting the railways with an absurdly complicated problem of shock protection? One way of solving the problem is to tie the component concerned in a strong canvas bag and then suspend the bag by soft springs inside a wooden crate. If the crate is dropped accidentally then the system within it can accept quite large deformations because of the softness of the suspension springs, and a remarkable degree of protection is found to be obtained in this way. When they use this idea, engineers alter the quantities that have been mentioned in connection with the apparatus of fig. 81; in particular, the system is given some very low frequencies of free oscillation (in which the springs within the crate are distorted and there is little distortion elsewhere).

Another common way of transporting delicate components is to pack them suitably in a readily deformable material, such as foam plastic. The effectiveness of this approach can readily be demonstrated. Thus an egg or an electric light bulb can be packed into expanded polystyrene and, when suitably done up, hurled at a wall without suffering ill effects. A more remarkable demonstration, however, is to be had from a dropping weight. If a 2 kg weight is dropped from a height of about 1.5 m it will produce a nasty dent in a piece of thick wood and it is obvious that anyone who rested his hand on a table under the falling weight would be asking for trouble. The weight can perfectly well be dropped on the hand, however, provided the hand is resting on a block of foam rubber, say about 10 cm thick. It is not even painful. In this little experiment, we not only change the duration of the impulsive loading and the first natural frequency of the system that is struck, but we also change the damping of the system quite substantially. Here, then, is another factor that might have to be allowed for in analysis.

Complicated vibrations

Fair moon, to thee I sing
Bright regent of the heavens
Say, why is everything
Either at sixes or at sevens?

It has only been possible to place vibrations in categories – free, forced, self-excited, and so on – because most vibrating systems exhibit certain properties. To a great extent, we have used the properties as hooks upon which to hang our discussion. There are, however, vibrations which do not fall into these nice watertight compartments and we must now see how this comes about. Very briefly, it is either because the systems that do the vibrating themselves vary as time goes on, or else their geometrical form varies significantly during their vibration. When a child 'works himself up' on a swing, for instance, he does it by systematically changing the distribution of his own mass, raising and extending his legs and leaning forwards and backwards.

6.1 Constant and varying characteristics

The natural frequency of a pendulum is pretty well independent of amplitude, provided the amplitude is not large. By timing, say, 50 swings and dividing the result by 50, we can determine the periodic time and hence the frequency with good accuracy. If the pendulum*

---

* A bicycle wheel with a weight attached to its rim serves well for this demonstration.

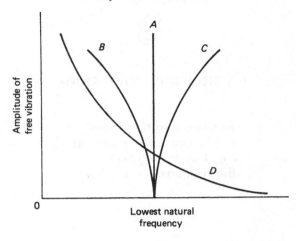

Fig. 85

is started from a small angle of inclination (up to about 30 degrees), very little difference can be found in the result, whatever the starting angle. But if the starting angle is increased to 60 degrees or more, then we discover that the frequency is diminished as the angle is increased.

This can be expressed in the form of a very simple graph. The line *A* in fig. 85 represents the prediction of simple theory. It tells us that the natural frequency of free oscillation of a simple system is independent of the amplitude. It is the sort of prediction that we have made throughout this book. But now we find, with our simple pendulum, that the line will bend over as in curve *B*.

Curve *B* bends over to the left: it is quite easy to find systems for which the amplitude–frequency curve will bend over to the right (as in curve *C*). This is so, for instance, with the flexural vibration of the loaded strip shown in fig. 86, since higher amplitudes of bending motion cause more and more limitation of the effective free length of the spring strip.

These simple systems afford an excellent example of the way in which the theory of vibration is used by engineers. A theory which predicts a curve like *A* becomes less accurate as the vibration amplitude is made larger. In cases of this sort, an engineer is not normally worried by the lack of accuracy. Generally speaking, his problems become somewhat academic if the oscillations become so

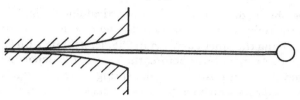

Fig. 86

violent that this begins to matter. It is fair to say that, to a mechanical or structural engineer, the difficulties associated with departures such as those of curves B and C from curve A are often more imagined than real. This is not always so, however, and in some of nature's oscillations varying characteristics are the general rule. A physiologist would not expect the heart to be an oscillating device with constant characteristics.

Let us now take up another problem of this type. Here, though, the frequency of even a modest oscillation is quite different from that of a very small one. If a glass bar whose cross-section is as shown in fig. 87 is made to rock backwards and forwards on a hard surface, first on one fulcrum and then on the other, its frequency is noticeably *less* in a free vibration of moderate amplitude than it is when the amplitude is very small. If the bar is tilted slightly and is left to rock, its frequency gradually increases as its amplitude dies away. The appropriate curve in fig. 85 is therefore of the form D.

Let us now examine why the curves A, B, C and D in fig. 85 differ. If a mass is suspended by a simple spring (or, better, a rubber cord) and is made to oscillate vertically, then a curve will be found which is a very close approximation to A. If there were no friction present, the oscillation would be a truly sinusoidal one. And if the spring were to be replaced with a stiffer one, then a similar curve A

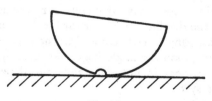

Fig. 87

would be obtained, but it would be a vertical line in fig. 85 further along to the right; so much for two different, *constant*, stiffnesses.

In explaining the other curves in fig. 85, let us start with C, as it is the simplest. As mentioned before, the greater the departure of the mass from its mean position (in the spring system of fig. 86), the stiffer is the spring which tries to push it back. It is as if the system with the bob in its mean position is a different system from that when the bob is displaced, for the stiffness is different – and the stiffness partly determines the natural frequency. The higher the stiffness is, the greater will the frequency be. A close examination of the motion would reveal that, even in the absence of damping, the waveform of the oscillation is not quite sinusoidal. This is also true of the large swings of the pendulum and of the rocking of the glass bar.

Turning now to the system of fig. 87, we find that it is not so much the stiffness that varies during a cycle of oscillation as the system itself. In one half-cycle the left-hand fulcrum is used and in the other the right, so it is certainly true that the characteristics of the system are not constant.

Coming back now to the pendulum with its curve B, we find that the explanation for this bending-over effect is a little more subtle. A simple way of looking at the matter, though, is to note that, when the amplitude is large, the distortions of the system are no longer 'small'. That is to say, the displacement of the bob is of the same order of magnitude as the size of the whole system.

The point here is that, when we say that the pendulum is oscillating, we are referring to the variation of the angle that the pendulum makes with the vertical, and we must therefore discuss the 'stiffness' of the system in terms of that angle. It happens that the force which tends to restore the pendulum to its vertical equilibrium position is dependent, not so much on the angle between the pendulum and the vertical as on its sine. For small values of this angle the sine and the angle itself are approximately equal, so that the system has effectively a restoring force that is proportional to the displacement. But for larger inclinations of the pendulum the restoring force becomes less than what it would be if it were dependent upon the angle itself.* So here again we have a system whose stiffness is not constant, but depends upon the distortion of the system. ·

* It will be remembered that, for positive values of θ, $\sin \theta$ is less than θ.

In the examples that we have mentioned so far, the stiffness of the system has been the thing that varies. Indeed springs are sometimes made in special forms so as to have stiffnesses which vary with distortion. In principle there is no reason why the mass should not vary instead of the stiffness, though technically this is less important. Finally, every normal system will have damping in some form and this too can be either constant or variable. It will be remembered that we limited the motion of the bouncing bar in fig. 56 by putting oil in the cups surrounding the lower springs. By this means the damping effect was increased *disproportionately* for large amplitudes. In a perfectly valid sense, then, the system was not the same for large amplitudes as it was for small ones.

6.2 The possession of constant characteristics

The amount and distribution of mass, damping and stiffness that is to be attributed to a system with constant characteristics (in a theoretical analysis of its vibration) is *fixed*. We do not contemplate, for instance, a spring stiffness that is dependent upon the time of day or on amplitude of motion. This is another way of saying that the restoring force between any two points of a system due to stiffness is always proportional to their relative displacement. Again, the only sort of friction that is allowed is the sort of friction that arises in a treacle of constant stickiness. This means that the damping forces resisting relative motion of any two points of a system are taken to be proportional to the relative velocity of those points. We do not allow a friction force which changes discontinuously with direction of motion, as in ordinary dry friction.

If this assumption of constant characteristics is made, then certain general properties of systems emerge, and it is these properties that we have previously taken for granted. Thus, the systems have inherent natural frequencies and principal modes of free vibration.

If such a system is subjected to sinusoidal excitation, its response also has a sinusoidal form (though it is generally out of phase), and it has the driving frequency. In this forced motion, the form of the fluctuating distortion depends only upon the characteristics of the system and upon the driving frequency; the *intensity* of the motion (for a fixed form of excitation) depends only upon the intensity of the stimulation. Moreover, the motion is perfectly stable. If the applied force is periodic, and not merely sinusoidal, then the force can be split up into sinusoidal components. Each component produces a motion and these are superimposed on each other to give the total response.

The position is more complicated if the system is non-conservative. Modal shapes may be more complex, sinusoidal forced motions in resonance modes may display phase differences, free vibration may grow or subside. And if free vibration grows it will gradually become infinitely large since the theory cannot account for limitation of amplitude, as we mentioned in section 4.3. But it still would not be true to say that results cease to conform to well-defined general rules.

The study of vibration, like that of mechanics, must be based on experiment. But, as we have already seen, some vibrations are too dangerous to be allowed, so that their onset has to be predicted if possible. This confers great importance on calculations, and for this reason the theoretical side of vibration analysis has become very highly developed by engineers. It is often necessary to calculate natural frequencies, modes, responses, critical speeds and conditions for instability. Generally speaking, significant headway can only be made in the calculations of a practical structure on the assumption of constant characteristics.

One of the drawbacks of vibration analysis, even with this assumption of constant characteristics, is the extraordinarily large volume of data which has to be marshalled for certain systems. An adequate representation of an aeroplane's motion may well require the use of, say, 50 mass numbers, 50 damping numbers and 50 stiffness numbers for any speed of flight and air density. The question of book-keeping therefore arises and electronic digital computers are used for making the calculations. The length of the sums makes them quite unsuitable for human calculators.

There is one particular branch of pure mathematics, namely matrix algebra, which lends itself to this type of work particularly well. Because of this, the mathematical theory of matrices has been very greatly extended since the late 1930s and one of the best-known standard works on matrix theory was written by three engineers (at the instigation of the Aeronautical Research Council) whose main interests lay in the flutter of aeroplanes.

Calculations of the type just mentioned are very common and are vitally important in engineering. The assumption of constant characteristics is perfectly adequate for most practical purposes and it would be foolish to undertake the far greater mathematical problems of varying characteristics for any but the most pressing need.

The trouble with the theory, however, is one which always arises with any theory. It is not just a question of numerical accuracy, but, rather, that we can never be sure that we have not approximated some vital matter right out of existence. It is probably true that there is no such thing as a system which has *absolutely* constant characteristics and there are many which do not have them even approximately. The question therefore arises as to whether or not we lose anything vital if we assume constant characteristics. We must now concentrate on that behaviour of systems with varying characteristics which is completely overlooked by a theory in which this assumption of constant characteristics is made.

If none of the general features that have been mentioned can be taken for granted, our discussion will clearly be a great deal more complex. We can no longer afford to make a frontal attack and we shall have to limit ourselves to very simple examples. If we start thinking about questions of vibration 'shape', for instance, we shall soon find ourselves in trouble. Accordingly, we shall now examine a few simple engineering vibration problems in which unexpected phenomena are observed. It would be quite impossible, with the present state of knowledge, to give a systematic account of this aspect of vibration.

6.3 Dry friction

We have alluded once or twice to the effects of dry friction, and have already pointed out that, strictly, this form of damping should be identified with systems possessing varying characteristics. This is because the friction force changes abruptly in direction as the direction of sliding changes (see fig. 63), and this property frequently sets up a form of self-excitation. In fact, whereas viscous friction forces almost invariably tend to suppress self-excitation, *dry* friction may actually cause it, as we have seen with the squeaking of hinges and the motion of violin strings.

Sometimes, a bicycle will shudder when the front brake is applied. The necessary source of energy is to be found in the motion of the bicycle itself, and the mechanism by which this energy is

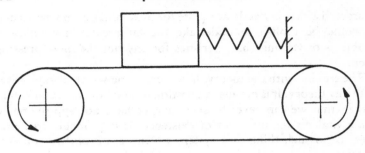

Fig. 88

changed into a vibration can only be found by studying the surface
physics of the brake block and the rim. (The shuddering can, in
fact, be reduced greatly by slightly altering the position of the brake
blocks.) It would not be easy to find a really accurate explanation
of the bicycle shudder or of the motion of a violin string. Some
progress may be made, however, with the apparatus shown in fig.
88. A spring-restrained wooden block rests on a continuous
sandpaper belt and, when the belt is made to go round, the block is
seen to shudder. The motion takes place because the friction
between the body and the belt is greater for small slipping velocities
than it is for large ones. This is illustrated by the portions of curve C
in fig. 63 which lie near the vertical axis. The velocity of the belt is
always greater than the fluctuating velocity of the block so that the
direction of sliding is always the same. But while the block moves in
the direction of the belt the velocity of slip is smaller than it is when
the block moves in the opposite direction. The friction force exerted
upon the block thus has a fluctuating component and the direction
of this fluctuating component always coincides with the direction
of velocity of vibration. We therefore have the ingredients of
unstable vibration.

Fig. 89(a) shows a cross-section of a shaft running loosely in a
bearing that has been allowed to become dry. The direction of
rotation of the shaft is shown by the arrow. Suppose that, while it is
running, the shaft becomes displaced and touches the side of the
bearing. The effect of the force P which now acts at the surface
(shown in fig. 89(b)) causes the shaft to roll round the bearing, for
the direction of P is always the direction of the tangent at the point

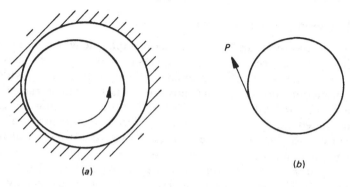

Fig. 89

of contact. If the speed is high enough, a violent vibration is set up. It can be quelled by the application of a little oil.

Self-excitation due to dry friction occurs in many forms. Fig. 90 shows a simple roller running on coned surfaces (as railway trains do to preserve a centring action). When it runs slowly down inclined rails it behaves as a civilized coned roller should; if put on straight it rolls down straight, but if put on cocked it will slowly zig-zag down the track. If the slope of the track is increased so that the speed increases sufficiently, however, the zig-zag motion builds up and the oscillation of the roller becomes violent. The explanation of this is to be found in the skidding which takes place at the rail-contacts and which is governed by the 'dry friction' law.

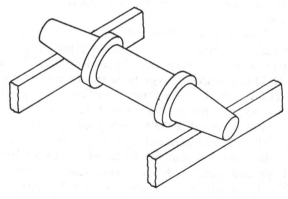

Fig. 90

6.4 Limitation of self-excitation

The variations of natural frequency with amplitude which we noted at the beginning of this chapter furnish us with a very simple example of the effects of varying characteristics. In order to widen our discussion, let us go back now to a matter that is raised in section 4.3. It will be remembered that the bouncing of the wind-excited bar of fig. 56 was limited by the introduction of oil in the cups at the bottom of the springs. A large deformation of the springs then means that more turns enter the oil so that the damping of the system becomes disproportionately larger for larger amplitudes. The self-excited oscillation thus becomes somewhat like that indicated in fig. 62. The motion grows from nothing and achieves a steady amplitude; moreover the resulting waveform is more or less sinusoidal (though not quite).

If the bar is released from the extreme position against its frame while the fan is running, the vibration does not grow but diminishes to the same final form. The motion is said to occur in a 'limit cycle'. A close examination of the motion would reveal that it occurs nearly at the natural frequency of bouncing and that, although it is not strictly sinusoidal, it is nearly so.

When there is oil in the lower cups of the system of fig. 56, the measure of damping will not depend upon amplitude in any discontinuous way but will vary steadily. Up to a certain amplitude, there is in effect a negative damping provided by the wind flowing past the bar. But as the amplitude becomes greater than this, the damping becomes positive – the effect of the oil exceeding that of the air flow.

While the bouncing bar is vibrating in its limit cycle it possesses a certain amount of energy. And, since the amplitude is constant, the average amount of this energy remains constant. Aside from this, however, the bar *transfers* energy from the airstream to the oil in the cups. The amount of energy transferred in this way per cycle is small compared with the average energy stored in the oscillation and it is for this reason that the bouncing motion resembles a free vibration of the bar. This is not always so, however, and there is an important class of vibrations in which the energy exchanged per cycle is large compared with the energy stored in the vibrating system. This is the sort of oscillation that can be heard when one rubs a finger over an inflated toy balloon. The limit cycle is then jerky and the frequency is determined by the physical process of the energy exchange and not by the dynamic character of the oscillatory system. Motion of this sort is given a special name – it is a 'relaxation oscillation', and it may be recognized by its characteristically angular waveform.* Some aspects of machine-tool chatter are thought to be explainable in terms of the theory of relaxation oscillations.

Vibration in limit cycles of one sort or another is very common. The squeaking of a door hinge and the swaying of a railway coach are two examples of it, both being self-excited motions that grow until they become controlled by some limiting process. Much of the importance of these limit cycles lies in the fact that they are often all that is easily observable, the initial self-excitation being soon controlled and modified by the limiting process; this is true, for instance, with railway coach oscillations. But unfortunately the subject of limit cycles is also a complicated one, so that further development of our introduction to their properties would be heavy going. Let us, therefore, leave this topic – merely observing that they constitute another important subject for research.

6.5 Time-dependent stiffness

We have begun to see that curious effects can arise from variations of damping. The damping characteristics of the systems mentioned in the previous section were dependent upon the distortions of the systems concerned. It is quite common for the *stiffnesses* of systems to vary perceptibly during vibration – indeed one or two examples of this were described in the opening section of this chapter. But now let us discuss stiffnesses that are not dependent upon displacement (or distortion) so much as they are dependent upon time. In fact the particular system that we shall examine has, or has not, got a time-dependent stiffness according to our way of looking at it!

The alternator rotor shown in fig. 41(a) is, in effect, a rotating magnet with a north and south pole. These poles are on diametrically opposite sides of the cross-section and are separated from each

---

* The term 'relaxation oscillation' is sometimes used in a broader sense. It is used, for instance, to describe *any* oscillations that have a more or less angular waveform, and in particular a 'sawtooth' waveform like that of the vertical motion of a yo-yo. Even the oscillations of a clock-escapement have been described as relaxation oscillations.

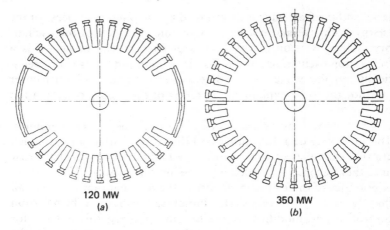

120 MW
(a)

350 MW
(b)

Fig. 91

other by longitudinal slots cut in the metal. The slots carry conductors which form closed circuits such that the current carried by them produces the magnetic field. Unfortunately it is not possible to obtain a magnetic field that is intense enough by simply rotating a permanent magnet.

Fig. 91(a) shows a diagram of the cross-section of a 120 megawatt alternator shaft near its middle after the slots have been cut and before the conductors have been placed in them. The conductors are carefully insulated and packed into the slots, and they are prevented from flying out when the shaft rotates by covers placed in the mouths of the slots. These conductors are heavy for they have to carry large currents.* At the ends of the slots, where the conductors have to be turned back on themselves so as to run down other slots, the conductors are prevented from flying out by 'end bells', the design and construction of which require great skill.

The central longitudinal plane of the poles is effectively a plane of maximum flexural stiffness of the rotor. The perpendicular plane is greatly weakened by the slots in the rotor so that it is a plane of minimum stiffness. As we shall see, rotating a shaft of this type at high speed can present some difficulties – and may even be dangerous. The difficulty arises from this discrepancy between the two bending stiffnesses. The shaft shown in fig. 41 is a 350 megawatt

* The rotor to which fig. 41 relates will carry about 29 000 amperes per slot.

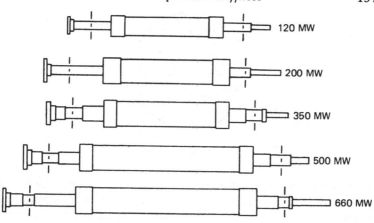

Fig. 92

rotor and it is not made in exactly the form just described. Its cross-section is like that shown in fig. 91(*b*) in which it will be seen that slots have been cut, not only for the conductors, but also in the pole faces. In order that the intensity of the magnetic field produced by the current in the conductors shall not be greatly reduced by this modification of the pole faces, however, the rectangular slots will later be filled with bars of steel. Let us now try to find out why one should wish to go to the length of slotting the pole faces and then filling the holes up again with bars of steel.

The first point to notice is that, generally speaking, modern alternators of higher output have longer rotors. (They must be larger and they cannot be fatter because of the high centrifugal stresses in a spinning shaft which tend to burst it.) Fig. 92 shows the relative sizes of rotors that have been used for these machines since the end of World War II. The shafts are becoming decidedly slender, and we have seen what this means – their critical speeds have been depressed and, dynamically, they are much more 'touchy' than they were in days gone by. In other words, engineers have been forced to consider some of the less familiar aspects of their behaviour.

Suppose that a thin steel shaft is used in a piece of apparatus like that of fig. 40, but that this time the shaft has flats on it so that its cross-section is as shown in fig. 93. Instead of the previous

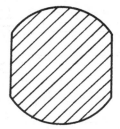

Fig. 93

apparatus, however, it will now be necessary to use a rig which can be stood on one end so that the shaft may be either horizontal or vertical while it rotates. When the shaft is securely mounted in its bearings it is found to possess two low natural frequencies corresponding to the stiff and weak planes of the shaft.

When the shaft is run up to speed while in the vertical position, its behaviour differs somewhat from that of the circular shaft. It runs quietly in its bearings until the lower of the first pair of natural frequencies is approached. As we should expect, perhaps, a resonant forced motion is produced at this speed (caused by a slight unbalance of the shaft, or a slight bend in it, or both) and the shaft starts rattling against its guard. If the speed is increased the rattling does not cease until the shaft has passed through the second of the lowest pair of natural frequencies (corresponding to the stiffer plane). There is a range of instability between these two rotational speeds and only outside this range will the shaft run quietly. This state of affairs is indicated in fig. 94.

Slotting of the pole faces of large turbo-alternators is practised only partly from the fear of instability between pairs of natural frequencies. In fact it is found that this dual rotating stiffness gives rise to other vibrations of the machine.

Before proceeding, however, let us think for a moment about the vertical shaft whose behaviour is summed up in the curve of fig. 94. There is nothing very unusual in the results so far – indeed, it would be surprising if there were. For, while a stationary observer would assert that the stiffness of the shaft varies through two complete cycles for every revolution of the shaft, an observer who rotated with the shaft would affirm that there is no varying characteristic. The system, as we have seen, displays both resonant

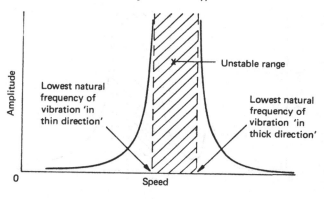

Fig. 94

forced motions (due to small inaccuracies of manufacture) and a range of (unlimited) unstable free motion. It happens, moreover, that the forced motion has the interesting feature of being sensitively dependent on the angle between the plane of the shaft's defect (of bend or unbalance) and its plane of maximum stiffness.

The demonstration shaft whose section is shown in fig. 93 does not behave in quite the same way if it is run in the horizontal position. It now performs an *extra* resonant motion at half the average of the previous resonant speeds. In the vertical position the weight of the shaft acts along the axis of rotation, while in the horizontal it acts across the axis.

Suppose that the weight *does* act across the shaft. The gravity effect is seen as an unvarying force acting in the vertical plane by the stationary observer (who, it will be recalled, sees the fluctuating stiffness). But the rotating observer sees the gravity effect as a steady transverse force which rotates (backwards) at the speed of rotation, though he still sees a shaft with constant characteristics. The plane of the 'rotating gravity' force coincides with a plane of maximum stiffness twice per revolution of the shaft, and the 'half speed' resonant hump indicated in fig. 95 is the result.

There is a simple modification that we may make to our apparatus which will force even our astute rotating observer to admit that the system has varying characteristics. This is to place a pair of transverse guides close up to the shaft so that the flexural vibration is confined to a single plane. Whichever way we look at

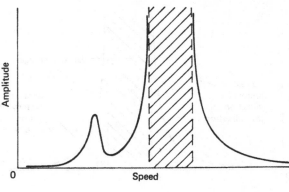

Fig. 95

the shaft now we conclude that vibration in the permitted direction must be determined by a stiffness which varies through two complete cycles during each revolution of the shaft. There is no escape from a much more difficult form of analysis now, whose predictions turn out to be much more complex than those indicated in figs. 94 or 95.

6.6 Concluding remarks

The oscillation of systems possessing variable characteristics is a vast and complicated subject. It has been introduced here through particularly simple systems having (*a*) displacement-dependent stiffness, (*b*) displacement-dependent damping and (*c*) time-dependent stiffness. This only scratches the surface of the subject, though, and there are all sorts of curious phenomena which become apparent in investigations.*

This degree of complication forces one to discuss the motion of systems having variable characteristics in such a way as to take each problem separately. It is no longer easy to categorize oscillations. As regards mechanical engineering, however, there are one or two general observations which are, perhaps, appropriate.

With a system having constant characteristics, a harmonic excitation produces a harmonic response. This is not necessarily so with a

---

* In mechanical systems it is usually the damping and the stiffness characteristics which are variable, though there are some problems in which the mass characteristics change, particularly in connection with reciprocating machinery.

system having variable characteristics, and the response may contain contributions having frequencies which are multiples or submultiples of the excitation frequency. Moreover it is possible for one of these components to be brought to resonance by the sinusoidal input and so magnified dangerously. It is also possible for a sinusodial excitation to produce a response which may be one of two – the one actually chosen being determined by the previous history of the motion. Yet again, it is possible for a forced motion to give rise to regions of 'unstable forced vibration' – which means that the forced motion does not actually occur! In fact, the very meaning of the word 'unstable' requires very careful definition indeed in this subject. Self-excitation may occur and produce motions of *limited* amplitude. These motions may be of nearly sinusoidal form or may be something very remote therefrom.

In engineering it is usually permissible to make the assumption of constant characteristics. Indeed, engineers sometimes make a virtue of necessity in this respect. A resonance test of an aeroplane, in which the airframe is shaken while it is on the ground and the motion is observed at various points of the structure, does not produce results which are entirely compatible with theory based on constant characteristics. But to make an allowance for the variation of stiffness and of damping within an airframe would be an incredibly difficult thing to do.

While all this is true, the engineer has to remember that, in some problems, the variation of characteristics is an essential feature of the system with which he is concerned. It would be pointless, for example, to expect to conduct a successful analysis of something like a clock escapement on the assumption of constant characteristics.

In this book we have seen how an engineer regards mechanical vibration. To him, a vibration problem is something which usually demands serious study, and often he can disregard it only at his – or someone else's – peril.

While an engineer may experience extreme difficulty in contending with man-made structures and machines, he has only to think about some of nature's oscillations to realize what wonderful, complicated things oscillatory systems can be. No engineer ever managed to design anything so marvellous, for instance, as that remarkable vibrating machine, the heart. There is no rotating machinery in the human body so that any device which is to serve as a pump has to be a reciprocating one. The heart operates for a lifetime and the maintenance that it demands is supplied (at least in part) by the results of its own action. Strictly speaking, it is not the same heart which goes on functioning year in and year out, for the process of maintenance involves gradual modification of the oscillating device itself.

It is perfectly obvious, of course, that the harder one looks at *any* physical process the more complex it appears. The art of the engineer lies (at least partly) in knowing when to stop peering at things and to start 'getting on with it'.

Index